Drugs and the Lung

ETTORE MAJORANA INTERNATIONAL SCIENCE SERIES
Series Editor:
Antonino Zichichi
European Physical Society
Geneva, Switzerland

(LIFE SCIENCES)

Recent volumes in the series

A Continuation Order Plan is available for this series. A continuation order will bring delivery of each new volume immediately upon publication. Volumes are billed only upon actual shipment. For further information please contact the publisher.

Drugs and the Lung

Edited by

Gordon Cumming

The Midhurst Medical Research Institute
Midhurst, West Sussex, United Kingdom

and

Giovanni Bonsignore

University of Palermo
Palermo, Italy

Springer Science+Business Media, LLC

Library of Congress Cataloging in Publication Data

School of Thoracic Medicine (6th: 1982: Ettore Majorana School of International
    Scientific Culture)
    Drugs and the lung.

    (Ettore Majorana international science series. Life sciences; 14)
    "Proceedings of the Sixth School of Thoracic Medicine, held June 1982 at the Ettore
Majorana School of International Scientific Culture"—T.p. verso.
    Includes bibliographical references and index.
    1. Respiratory agents—Congresses. 2. Lungs—Effect of drugs on—Congresses. I.
Cumming, Gordon. II. Bonsignore, G. III. Title. IV. Series. [DNLM: 1. Lung diseases—
Drug therapy—Congresses. 2. Lung—drug effects—Congresses. W1 ET712M v. 14]
RM388.S36   1982   615'.72   83-24593

Proceedings of the Sixth School of Thoracic Medicine, held June 1982 at the
Ettore Majorana School of International Scientific Culture

© 1984 Springer Science+Business Media New York
Originally published by Plenum Press, New York in 1984
Softcover reprint of the hardcover 1st edition 1984

ISBN 978-1-4757-1271-1   ISBN 978-1-4757-1269-8 (eBook)
DOI 10.1007/978-1-4757-1269-8

PREFACE

This volume records the proceedings at the Sixth School of Thoracic Medicine held at the Ettore Majorana School of International Scientific Culture in June 1982.

Foregathered there were a heterogeneous group comprising clinicians, pharmacologists, pathologists, ultra-microscopists, biochemists and immunologists and they presented the eighteen papers seen in the contents list. The discussion which followed each paper was faithfully recorded (and where necessary translated) and may be found after each author's presentation. This free discussion is perhaps the most valuable part of the School of Thoracic Medicine, and most clearly defines the present boundaries of knowledge, and the directions in which enquiry is being pursued.

The collaboration of many people made the production of this book possible – for translation and the discussion typescript Miss Guiliana de Ferio; for the final typing and layout Miss Corinne Wade. The illustrations have been dealt with where necessary by Mr. John Griffiths and the production of the book was done at The Midhurst Medical Research Institute prior to its delivery to Plenum Press.

G. Cumming

G. Bonsignore

CONTENTS

DRUGS AND THE LUNG

CENTRAL AND PERIPHERAL SITES OF ACTION

Sheila Jennett

Institute of Physiology

University of Glasgow, Scotland

The links between drugs and the lungs, and the distinction between central and peripheral sites of action, may be explored under three main headings:

1) Effects of systemically administered drugs on respiratory function itself acting peripherally or centrally to alter breathing pattern, to alter ventilation, or to affect oxygen transfer.

II) Effects of systemically administered drugs on the non-respiratory functions of the lungs: agents acting either systemically or locally, may alter the defence functions or the metabolic functions of the lungs.

III) Effects on the lungs themselves, or in the rest of the body, of drugs which are administered via the lungs.

This introductory overview will present a framework within which the often complex and multiple effects may be considered: the different specialist contributions which follow will enlarge on particular aspects within this scheme. The aim is to show by what mechanisms the physiological function may be altered, rather than to detail particular pharmacological effects.

I. Effects on respiratory function

It is useful to subdivide effects on respiratory function under the headings of those on breathing pattern, those altering the ventilation, and those affecting oxygen transfer.

These may, of course, overlap; it is for example unusual for a drug to depress ventilation without also altering breathing pattern, though the converse is common. The distinctions are useful ones, if a pharmacological action is to be clearly defined and if a clinical effect is to be assessed.

Breathing pattern is normally essentially regular and this regularity depends on the normal functioning of the complex of respiratory neurones in the brain stem and on their generation of a rhythmic output, modulated by various inputs. Central computation results in a pattern of tidal volume and respiratory frequency which for the individual involves the least possible muscular work. The controlling mechanism is provided with information concerning the state of inflation and deflation of the lungs (though there is evidence that this vagally mediated afferent input may play little part in modulating resting breathing in adult man) and with information on the tension being developed in respiratory muscles as the chest cage is expanded. Receptors in the lungs and in the muscle spindles, their afferent pathways and integrative function both at the spinal segmental level and in the brain stem, are involved in maintaining the normal pattern of breathing. The pattern in a subject who is at rest and undisturbed, may be regarded as a function of the control centres and the 'mechanical feedback' information. (Fig. 1.)

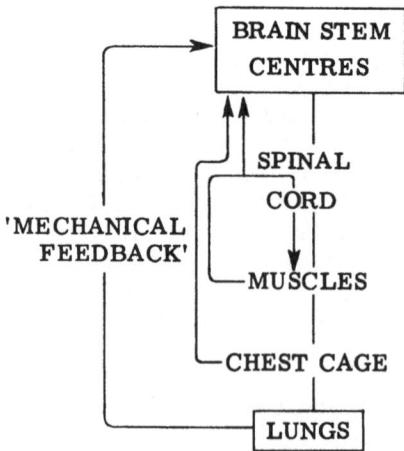

Figure 1

The rhythmic respiratory cycle involves the control not only of the muscles of inspiration and expiration, but also of the skeletal muscles of the upper respiratory tract and of the smooth muscle of the bronchial tree: the continual modulation, for example, of laryngeal muscle activity and of bronchial diameter with inspiration and expiration. Thus alongside the somatic efferent rhythmic output, there is also an autonomic rhythmic output from brain to lungs. (Fig. 2).

Changes in pattern may involve change in the frequency/tidal volume relationship, with the pattern remaining regular, or they may involve interference with rhythmicity, causing irregularity. Various types of drug or other interference, at the several possible sites of action, have complex effects, but some may be singled out for description.

Increase in respiratory frequency - tachypnoea - is not necessarily associated with an increase in the minute volume of ventilation: it can occur quite independently, so that counting the number of breaths per minute is not a reliable guide to the actual ventilation. Halothane, for example, can increase frequency yet depress tidal volume. Another example of this dissociation is commonly seen in cases of acute brain damage, in which tachypnea is common, and may or may not be accompanied by true hyperventilation (Jennett 1982). Although the stimulus and site of action is not always clear, this phenomenon can sometimes arise directly from brain stem damage but is more usually attributable to altered afferent inputs, from the periphery. One of these inputs might be from the J-receptors in the lungs, which Paintal (1973 and personal communication) has associated particularly with tachypnoea; these receptors can be stimulated chemically, or by factos increasing pulmonary interstitial pressure. Tachypnoea is associated with increasing body temperature; however drugs which affect body temperature commonly have other diverse actions on respiration also (Cooper & Guenter, 1981). Tachypnoea therefore could be a feature of agents acting centrally or peripherally.

Examples of the opposite condition a slowing of breathing, which is not necessarily related to underventilation, may be seen with some of the opiate analgesics; irregularity and bradypnoea may occur, but tidal volume may increase sufficiently to compensate for the decreased frequency (e.g. Jennett, 1976, Fig. 1).

The pattern of frequency and tidal volume alter when the resistive load to breathing changes; it therefore follows that drugs which alter this load can lead indirectly to a

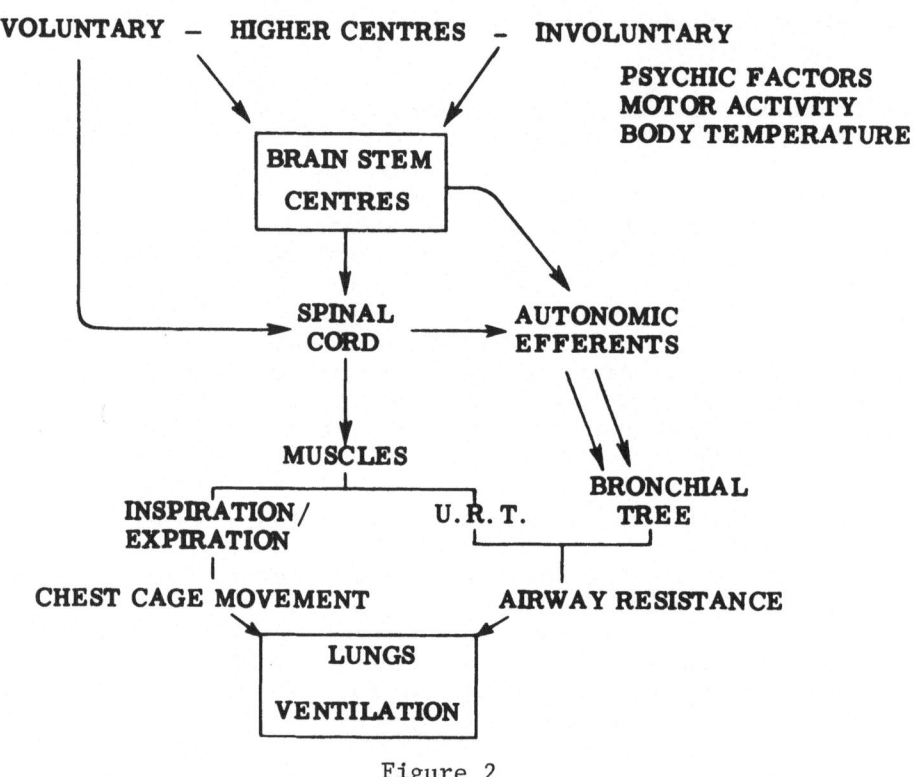

Figure 2

changing pattern for example if airway calibre is altered.
When the relationship is altered between the strength of
inspiratory muscle contraction and the lung expansion which it
achieves, the muscle spindle afferent information and the
spinal segmental integration are affected (Newsome Davis,
1970). Comparably, paresis of respiratory muscles, as for
instance by partial neuromuscular blockade, changes the
pattern; there is greater apparent effort resulting in less
effective contraction, and this alters the afferent
information.

In summary, drugs may alter the pattern of breathing, acting
centrally in the brain stem, or peripherally by affecting
receptors in the lungs or respiratory muscles, by altering the
load on those muscles or by weakening them. Such actions can
be independent of effects on the actual amount of the pulmonary
ventilation; but pattern changes are liable to be related to
alterations in the ventilation, which will be considered next.

Ventilation

Ventilation may be altered either appropriately, to match a
change in metabolic rate and therefore in requirement for gas
transfer; or inappropriately by agents which stimulate or
depress the respiratory control mechanism either acting
directly in the brain stem, or indirectly on the inputs which
affect it. These two types of change are distinguished by
whether or not the arterial PCO_2 is altered. Also an agent may
act indirectly to correct an existing disturbance of
ventilation: drugs which alleviate anxiety, for example, will
dimihish any hyperventilation which accompanies it, and this
effect must be distinguished from a respiratory depressant
action. Reticular activation during wakefulness accounts for a
significant drive on breathing, so that ventilation decreases
during sleep; hence any drug which is hypnotic will decrease
ventilation in this sense, and need not be considered
specifically depressant, unless the effect on ventilation is
greater than that attributable to sleep alone. Inputs from
somatosensory sources can stimulate breathing: so drugs which
relieve pain, may cause a decrease in ventilation, again
without being direct respiratory depressants.

Direct and reflex alterations in ventilation: effects on
ventilation which are not related to changes in metabolic
requirement, can be brought about by direct action on
respiratory neurones themselves, as for example by the
stimulant action of analeptics or the depressant action of
narcotics. These however are usually not specifically
respiratory actions, but are parts of generalised actions on

the central nervous system. More specific stimulation and depression could theoretically be brought about by action of a drug at any receptor site which can influence ventilation: a framework for considering such actions is shown in Fig. 3.

The receptors and afferent pathways which normally maintain the correct alveolar ventilation are mainly those concerned with the 'chemical feedback' information, which inform the control centres of the effectiveness of alveolar ventilation in terms of the arterial blood gas levels, and of cerebral extracellular acidity (Fig. 3).

The central chemosensitive mechanism involves sites near the ventral surface of the medulla, responsive to changes in extracellular acidity, which in turn is readily influenced by alterations in arterial PCO_2; pharmacological depression or stimulation of this mechanism would alter ventilation accordingly. By acting at this site, it would be possible to depress ventilation and the responsiveness to carbon dioxide, and yet leave the response of the respiratory neurones to other incoming stimuli unaffected. (It was indeed the demonstration of such a dissociation under some anaesthetic agents, that led to the prediction and search for a separate chemosensitive area).

There is little information about drugs which might have this particular site of action, because the respiratory depressant or excitatory effects of drugs have traditionally been tested by measurement almost entirely of the effect on the response to rising CO_2. Such tests cannot distinguish between depression of the respiratory neurones themselves, and depression of central chemosensitivity.

To reach either the central chemosensitive mechanism, or the brain stem neurones themselves, a drug has to pass the blood brain barrier.

Stimulation of the peripheral chemoreceptors mimics physiological stimulation by low oxygen tension, and results in hyperventilation with hypocapnia.

In man and in intact animals it is not possible to distinguish between central and peripheral stimulation, since a drug given systemically will reach both receptors, but experimentally the sites of action can be separated. For example, it was initially believed that doxapram was a central nervous stimulant with a selectively potent direct action on the brain stem respiratory neurones; but animal studies showed that most of the respiratory stimulant effect was accounted for

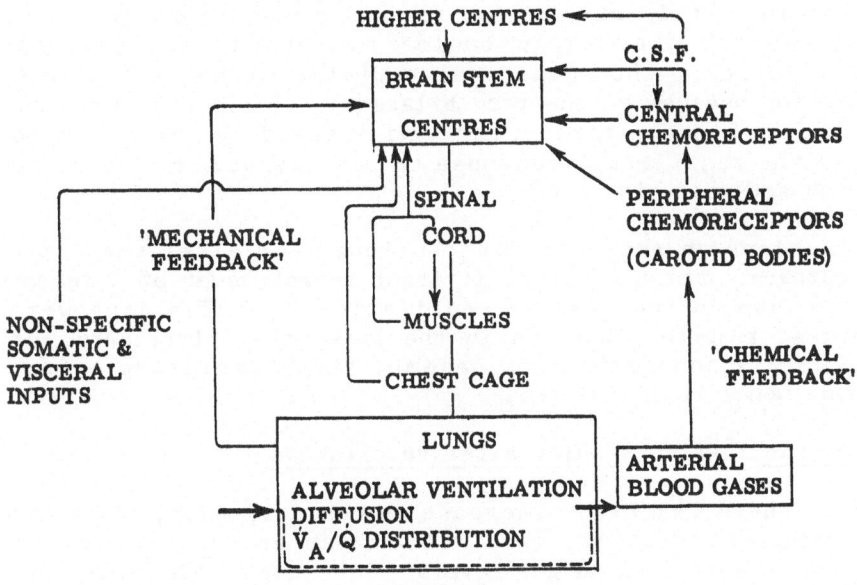

Figure 3

by its action on the carotid bodies, and was abolished by
section of their afferent pathways; only in large doses was
there a central stimulant effect (Hirsh & Wang, 1975). More
recently, almitrine has been shown to stimulate respiration via
these same receptors. (Laubie & Schmitt, 1980). The
distinction between central and peripheral chemoreceptor
stimulation is of importance in that the more specific
peripherally acting agent does not have the disadvantages of a
generalised analeptic. However, it must be borne in mind that
peripheral chemoreceptor stimulation does not affect only the
respiratory control system: there are also reflex
cardiovascular effects when the carotid and aortic bodies are
stimulated: an agent which stimulates respiration by this
route can be expected also to increase the arterial blood
pressure. It is of interest in this context that the carodid,
rather than the aortic bodies mediate reflex effects on
ventilation; the occasional patients without carotid body
function because of the rare bilateral carotid body tumours, or
arterial disease particularly if treated by endarterectomy,
lose the ventilatory response to low oxygen, and to any agent
which acts at this site.

Actions on receptors in the lungs can alter the drive to
breathing: stimulation of irritant receptors or of J receptors
can cause an increase in ventilation; so if a drug were to
depress receptor function in the presence of irritants, or of
pulmonary congestion and oedema, the stimulating effect of
these would be diminished.

Mechanical effects which alter ventilation

If there is a heavy increase in the work load, as by severe
bronchoconstriction, ventilation may become inadequate to
maintain normal blood gas levels; likewise if the work load is
normal but the respiratory muscles are weakened.

In summary there is a variety of ways in which drugs may act
to alter ventilation: true stimulation or depression is
recognised by a decrease or increase in arterial PCO_2, and may
be due to action directly on respiratory neurones, or on any of
the central or peripheral factors which normally influence
their output.

Oxygenation of blood in the lungs can be considered separately
from effects on ventilation. Changes in ventilation will of
course alter alveolar and arterial oxygen tension pari passu
with the opposite changes in carbon dioxide tension; but
disproportionate hypoxaemia is a separate consideration with
separate causes; it must be related to ventilation-perfusion

mismatch or diffusion defects, resulting effectively in some degree of venous admixture. Again, these functions can be altered by drugs or other factors acting either centrally or peripherally, and having effects either on the vasculature or on the airways.

Autonomic effects on oxygen transfer: it has been claimed that acute brain damage leads to pulmonary shunting, and also to congestion and oedema, by way of centrally generated autonomic effects on pulmonary vascular resistance; some workers have found evidence in experimental animals that there is protection from this by α -blocking drugs or by lung denervation (autotransplantation) (Moss et al 1973). On the other hand, there is evidence that sympathetic activity has a beneficial effect on ventilation-perfusion distribution: it may be necessary to prevent gravitational over-distension of vessels in the dependent regions particularly when total pulmonary blood flow increases (Szidon & Fishman 1969; Piene, 1976); also adrenergic receptors, and H1 and H2 histamine receptors, have been shown at least in some species to mediate constriction and dilatation of some part of the pulmonary vascular bed. This is a complex area with unresolved contradictions (see Jennett, 1982) and it can only be said that agents acting centrally to alter autonomic function, or sympathomimetic or parasympathomimetic agents (or their antagonists) acting peripherally in the lungs, are liable to have effects on ventilation-perfusion distribution, on the rate of formation of interstitial fluids, or on surfactant secretion, which are multiple and so unpredictable in their end-result. The only way to discover any resultant effect of a drug regime on oxygen transfer is to estimate the venous admixture. This is unlikely to be done routinely, since it properly entails obtaining samples of mixed venous blood; but the less invasive estimation of alveolar-arterial oxygen difference gives a guide to any changes which have taken place, if it is assumed that metabolic oxygen consumption and cardiac output have remained constant.

Other agents acting peripherally

Vasoactive agents reaching the lungs from the systemic circulation are known to be removed or catabolised in the lungs; any interference with the enzyme systems, probably mostly in capillary endothelial cells, which carry out this activity, could expose the vascular smooth muscle of pulmonary vessels to their action. Derangements of prostaglandin formation could likewise alter local blood flow.

A variety of drugs have been reported occasionally to have

effects attributable to increase in pulmonary vascular pressure, which could imply alterations in V_A/Q or in lung fluid; there may well be side effects of this type, on pulmonary pressure and regional flow, which go undetected, again because the necessary measurements are unlikely to be made. Increase in pulmonary arterial pressure does not of itself imply any impairment of gas exchange, but could be associated with uneven effects on the vasculature.

Vasoactive agents and the assessment of the effects of drugs on ventilation-perfusion distribution will be dealt with by later contributors.

II Effects on non respiratory functions

These can be classified as defensive and metabolic.

Defence function – consists essentially of prevention of drying and cooling, and of chemical, particulate or microbial contamination; it is mediated by mucosal and macrophage activity. It can be affected by centrally acting agents, in so far as the autonomic nervous system influences mucus secretion and upper respiratory air-conditioning, by variations in mucous membrane blood flow. Drugs acting systemically or locally can modify the nature, as well as the rate of secretion of mucus (Richardson & Phipps, 1981). Ciliary activity is depressed by a variety of agents including local anaesthesics in the respiratory tract, and such commonly encountered medicaments as atropine and codeine. Macrophage function can be disturbed by drugs which alter their cellular metabolism: by this mechanism antibacterial defences are impaired, experimentally in animal models, by a variety of agents (Gee & Khanwala, 1977).

Metabolic function

The lungs alter many substances reaching them from the systemic circulation, with the help of a variety of enzymes, probably mostly sited in the endothelium (Vane, 1976, 1968; Bakhle, 1976). The possibility exists that drugs interfering with or enhancing these enzyme actions may modify the extent to which vasoactive substances pass through to the arterial systemic circulation.

Drugs which have been administered systemically may themselves be taken up or bound in the lungs: there is a wide spectrum of pulmonary enzymes involved, in different types of cells (Hook and Bend, 1976).

Special attention is given to these topics in later sections.

Cholinergic mechanisms – an example of multiple effects.

As an illustration of the many ways in which drugs may be linked with the lungs, it is of interest to consider the involvement of cholinergic mechanisms, and therefore the potential effects of drugs which influence them.

The respiratory function itself is involved, with respect to the control of the <u>pattern of breathing and the ventilation</u> as follows:

There is experimental evidence that a group of brain stem <u>respiratory neurones</u> may have a cholinergic pacemaker type of function. Atropine does not have any noticeable effect on respiratory control, but of the anticholinesterases, physostigime crosses the blood brain barrier readily (neostigmine does not) and causes respiratory stimulation followed by failure with larger doses.

The <u>central chemosensitive mechanism</u> may depend on cholinergic synaptic transmission, accentuated by increasing extracellular acidity (Fukuda & Loeschcke, 1979), rather than on specialised receptor cells. This could be linked to the effect of hemicholiniums, which prevent endogenous acetyl choline formation, and which depress breathing, and diminish the response to CO_2 (Borison, 1981). The <u>peripheral chemoreceptors</u> have been thought to depend, for some component of the transduction of hypoxaemia, on a mechanism involving acetyl choline.

Peripheral factors affecting ventilation, involving cholinergic mechanisms include:

Effect on <u>airway resistance</u> by action on bronchial smooth muscle – relaxation by atropine, constriction by parasympathomimetics and anti-cholinesterases.

Effects on the <u>respiratory muscles</u> themselves – correction of the paresis of competitive neuromuscular blockade, by anticholinesterases, but eventual paralysis of respiratory muscles by toxic doses, on account of depolarisation block.

With respect to <u>oxygenation</u> in the lungs

In so far as this may be modified by any agent which can affect blood vessels and airways, and thence ventilation-perfusion distribution, the fact that there are muscarinic acetylcholine receptors in <u>pulmonary arterioles and venules,</u> as well as in bronchial smooth muscle, is potentially significant for gas exchange.

The action of cholinergic mechanisms in influencing bronchial secretions can indirectly affect alveolar ventilation and collapse.

The <u>non-respiratory functions</u> of the lungs are also influenced by cholinergic mechanisms:

<u>defence</u> function in that <u>airway secretions</u> are involved; also <u>ciliary motility</u> is affected (ciliary beat frequency has been shown to be increased by acetyl choline and depressed by atropine); lysozomal enzyme release in <u>macrophages</u> is stimulated by acetylcholine, which may relate to changes in phagocytic activity.

These potential effects on virtually every component of function related to the lungs, exemplify the complex possibilities when the several central and peripheral sites of action are considered.

To end with speculation – the areas of developing interest with respect to drugs and the lungs would seem to be the further elucidation of the roles, <u>centrally,</u> of the several neurotransmitters, and therefore the effects of agents which modify or mimic them; serotonin may be a tonic respiratory modulator; acetylcholine may be crucial in central chemosensitivity; endorphins are likely to be of significance in brain stem control (Bradley et al, 1976). <u>Peripherally</u> the possibility of influencing the metabolic function of pulmonary endothelium or macrophage enzyme systems, offers far-reaching possibilities both in lung disease and in its implications for the whole body.

REFERENCES

Bakhle, Y.S. (1976). The inactivation of endogenous amines in the lung. <u>Agents Actions</u> 6, 505–509.

Borison,H.L. (1981). Central nervous respiratory depressants. In: <u>Respiratory Pharmacology</u>, Ed: J. Widdicombe, pp 65–83 and 751–752. Pergamon Press Oxford.

Bradley,P.B., Briggs, I., Cayton,R.J.L.A. (1976). Effects of microiontophoretically applied methionine-enkephalin on single neurones in rat brain stem. Nature, 261, 425-426.

Cooper,K.E. & Guenter,C.A. (1981). The effect on pulmonary ventilation of drugs which influence body temperature. In: Respiratory Pharmacology, Ed: J. Widdicombe. Pergamon Press, Oxford. pp. 259-267.

Fukuda,Y. & Loeschcke,H.H. (1979). A cholinergic mechanism involved in the neuronal excitation by H^+ in the respiratory chemosensitive structures of the ventral medulla oblongata of rats in vitro. Pflugers Arch, 379, 125-135.

Gee,J.B.L. & Khandwala, A.S. (1977). Motility, transport and endocytosis in lung defense cells. In: Respiratory Defence Mechanisms Part II. Ed. J. D. Brain, D. F. Proctor & L. M. Reid. Marcel Dekker Inc. New York, pp 926-981.

Hirsh,K. & Wang,S.C. (1974). Selective respiratory stimulating action of doxapram compared to pentylenetetrazol. J. Pharmac. exp. Ther. 189, 1-11.

Hook,G.E.R. (1976). Pulmonary metabolism of xenobiotics. Life Sci., 18, 279-290.

Irivani,J. & Melville,G.N. (1981). Mucociliary function in the respiratory tract as influenced by physciochemical factors. In: Respiratory Pharmacology, Ed: J. Widdicombe. Pergamon Press, Oxford. pp 477-500.

Jennett,S. (1976). Method of studying the control of breathing in experimental animals and man. Pharmac. Ther. B. 2, 683-716.

Jennett,S. (1982). Pulmonary function in the head injured patient. In: Head Injury and Anaesthesia. Ed. W. Fitch and J. Barker. In press. Elsevier/North Holland, Amsterdam.

Laubie, M., Schmitt, H. (1980). Long-lasting hyperventilation induced by almitrine: evidence for a specific effect on carotid and thoracic chemoreceptors. Eur. J. Pharmac., 61, 125-136.

Moss,G., Staunton,C. & Stein,A. (1973). The centrineurogenic
    etiology of the acute respiratory distress syndromes.
    Amer. J. Surg., 126, 37-41.

Newsom Davis,J. (1970). Spinal control. In: The Respiratory
    Muscles. Ed. E. J. M. Campbell, E. Agostoni, J. Newsom
    Davis. Lloyd Luke, London. pp. 205-233.

Paintal,A.S. (1973). Vagal sensory receptors and their reflex
    effects. Physiol. Rev., 53, 159-227.

Piene,H. (1976). The influence of pulmonary blood flow rate on
    vascular input impedance and hydraulic power in the
    sympathetically and noradrenaline stimulated cat lung.
    Acta Physiol. Scand., 98, 44-53.

Richardson,P.S. & Phipps,R.J. (1981). The anatomy, physiology,
    pharmacology and pathology of tracheobronchial mucus
    secretion and the use of expectorant drugs in human
    disease. In: Respiratory Pharmacology. Ed. J. Widdicombe.
    Pergamon Press, Oxford. pp 437-475.

Szidon,J.P. & Fishman,A.P. (1969). Autonomic control of the
    pulmonary circulation. In: A. P. Fishman and H. H. Hecht
    (Eds). The pulmonary circulation and interstitial space.
    Univ. Chicago Press, Chic. and London, pp.239-265.

Vane,J.R. (1968). The alteration or removal of vasoactive
    substances by the pulmonary circulation. In:
    Importance of Fundamental Principles in Drug Evaluation.
    Raven Press, New York. pp. 217-235.

DISCUSSION

SPEAKER: JENNETT   **CHAIRMAN: CUMMING**

DENISON: In one of your slides you indicated that
            sympathetic activity could lead to a rise in
            capillary pressure and I suppose by that you mean
            veno-constriction, since obviously arterial
            constriction produces a reduction in capillary
            pressure. Is there good evidence that this
            actually occurs?

JENNETT: I am not sure that the evidence is good. The
            evidence that there is a sympathetic effect which
            leads to the complex of increased interstitial
            fluid and hypoxemia comes from rather complicated
            experiments. For example, one lung is denervated
            by auto-transplantation and this supposedly
            protects from the formation of oedema. I don't
            believe that there is very good evidence for veno-
            constriction in particular.

DENISON: You mentioned that the sympathetic system might
            influence the production of surfactant. Is there
            good evidence for that? That's an exciting idea
            and I just would like to know more about the
            evidence for it.

JENNETT: Yes, that evidence is from animal experiments
            where several things have been looked at in
            relation to sympathetic stimulation - both the
            lung extracts and also the mechanical behaviour in
            terms of the pressure - volume relationship before
            and during sympathetic stimulation. Dr. Shields
            tried, in my laboratory, to repeat this in pithed
            cats with appropriately localised stimulation of
            the sympathetic airflow but showed no change in
            mechanical behaviour which could suggest a change
            in surfactant secretion but there is a series of
            papers by Beckman & Associates where they seem to
            produce normally mediated pulmonary effects as a
            result of head trauma or more direct sympathetic
            stimulation.

DENISON: How do you see the role of the cholinergic system
            in the carotid bodies? What is your present image
            of how the carotid bodies work?

JENNETT: It has been controversial over the years whether
                or not there is a cholinergic component in the
                transduction process and I would not like to
                pronounce on the present state of knowledge about
                that.   But there has been demonstration of the
                appropriate enzymes in the carotid bodies.

CUMMING: Perhaps I could make a comment about the
                ventilation-perfusion ratios which you referred
                to.   The ventilation-perfusion ratio is the final
                result after the lung has done everything it can
                to maintain the milieu interieur.   It is possible
                to separate ventilation and perfusion: thus it is
                possible to have a marked defect in ventilation
                which is adapted to by a re-routing of the
                pulmonary blood flow.   This means the V_A/Q is
                substantially preserved in the face of a marked
                defect in the ventilation component: but any drug
                which is then administered, which in some way
                affects the vaso-active state of the pulmonary
                circulation, will cause some change in
                ventilation-perfusion ratios.   One or the other
                may be independently affected and have an affect
                on the PO_2.

DRUG RECEPTORS IN THE LUNG ESPECIALLY HISTAMINE RECEPTORS

Noemi Eiser

Department of Thoracic Medicine

New Cross Hospital, London

Drugs produce their actions by combining with receptors. A specific response is produced when an agonist binds to a receptor. In contrast, when an antagonist combines with that receptor, it produces either no response or a decrease in the expected response of the agonist. Pre-treatment with the antagonist will shift the dose-response curve of the agonist to the right. This is known as competitive inhibition. Thus, while agonists have affinity and efficacy, antagonists have affinity only. In many cases the precise nature and location of the receptor is unknown but the properties of established agonists and antagonists have been utilized to investigate receptors both in vitro and in vivo. In this paper I shall concentrate on the findings of recent research into histamine receptors in the lung and then consider, briefly, the evidence concerning cholinergic, adrenergic and noncholinergic, nonadrenergic receptors.

The effect of histamine on the airways varies widely between species. Thus, while in the rat, cat and ferret airways histamine produces either no response or bronchodilatation, in other animals, including sheep, horse, dog, guinea-pig and man, histamine is a powerful bronchoconstrictor. In man, the effects are particularly potent in asthmatics and this increased responsiveness to histamine is considered by some to be a diagnostic feature of asthma (Muittari, 1968). Even within the same species histamine may produce contrasting effects in different sites in the airways. For instance, in vitro histamine induces relaxation of the trachea and contraction of bronchi and lung strips in the rabbit, but relaxation of the bronchi and contraction of the trachea and lung strips in sheep. In 1966,

17

Ash and Schild first postulated the existance of two types of histamine receptor - the H_1-receptor whose actions could be prevented by mepyramine and the H_2-receptor whose effects were refractory to mepyramine block. Black and colleagues (1972) further defined these receptors when they developed specific agonists and antagonists to the H_2-receptor. These developments suggested the possibility that the differences in effect of histamine might be due to differences in the distribution of the receptors in the airways. Many studies using H_1- and H_2-receptor agonists and antagonists have confirmed this possibility, but unfortunately much of the data, which has been reviewed recently by Chakrin and Krell (1980) and Chand (1980), appears contradictory, due to differences in experimental techniques in the various in vitro studies. Nevertheless, it is clear that, in general, H_1-receptors mediate bronchoconstriction and predominate throughout the airways of most species, apart from the rat airways and the cat bronchi, where no H_1-receptors have been demonstrated and the ferret airways, where H_1-receptors do not predominate. It is thought that H_2-receptors are present in the trachea and bronchi of most species and usually mediate bronchodilatation. However, their presence has been determined by rather indirect methods - usually by demonstrating bronchodilatation when histamine is given in the presence of mepyramine and then by preventing that bronchodilatation by pre-treatment with H_2-receptor antagonists. With the exception of the guinea pig, H_2-receptors have not been demonstrated in parenchymal lung strips. In order to explain the unexpected findings in some of the studies, it has been suggested that there may be additional, atypical H_2-receptors in some sites. These receptors are similar to H_2-receptors since they are stimulated by histamine and 4-methyl histamine, but atypical since they are not protected by known H_2-receptor antagonists. They appear to be present in the trachea of the horse, rabbit and ferret and in the bronchi of cat and cattle.

To date only one in vitro study has addressed itself to the status of histamine receptors in human airways. Dunlop and Smith (1977), studying the effect of H_1- and H_2-receptor antagonists on antigen-induced contractions of sensitised human bronchus, reported the presence of bronchoconstricting H_1-receptors and bronchodilating H_2-receptors. Previously, Lichtenstein and Gillespie (1973) had demonstrated that H_2-receptor antagonists interfere with the autoregulation of IgE-mediated histamine release from human leukocytes. The implications from these two studies were that the administration of H_2-receptor antagonists to asthmatics might produce a dangerous enhancement of their bronchoconstriction. However, the validity of Dunlop and Smith's results is in some

doubt since they used very large doses of antagonists and did
not demonstrate dose-response relationships. In addition,
Kaliner (1978) was unable to demonstrate increased histamine
release by H_2-receptor antagonists from human lung in vitro.

A large literature has accumulated concerning the effect of
H_1-receptor antagonists on human bronchospasm in vivo. Many H_1-
receptor antagonists have been developed. However, since most
are relatively non-specific – having significant
anticholinergic, local anaesthetic and antiemetic properties,
their usefulness in the localisation of H_1-receptors is
somewhat limited. Nevertheless, certain facts have emerged.
When given parenterally, H_1-receptor antagonists decrease
bronchial tone in asthmatic, though not in normal subjects
(Herxheimer, 1949; Booij-Noord et al, 1970; Casterline and
Evans, 1977; Woenne et al., 1978; Nogrady et al, 1978; Popa,
1977; Eiser et al, 1981). For instance, thiazinamium 50 mg was
found to increase FEV_1 more than 2 puffs from an isoprenaline
aerosol (Booij-Noord et al, 1970), chlopheniramine i.v. to
bronchodilate to a comparable extent to aminophylline 5.5 mg/kg
(Popa, 1977) and inhaled clemastine and salbutamol to produce
similar effects in the airways of asthmatics recovering from
acute asthma (Nogrady et al, 1978). These data have prompted
speculation that there is a continuous outpouring of histamine
in asthmatics, producing a histamine tone. Unfortunately the
extent of the bronchodilating effect of these H_1-receptor
antagonists may be limited by local irritation if the drug is
inhaled (Nogrady et al, 1978), local histamine release (Church
and Gradidge, 1979) and by the difficulty of giving adequate
oral or i.v. doses because of the associated CNS depression,
which is present to some degree with all the H_1-receptor
antagonists.

The effects of H_1-receptors antagonists on induced
bronchospasm has been investigated repeatedly. The bronchial
effects of inhaled histamine have been significantly attenuated
by a variety of H_1-receptor antagonists, given by different
routes, to normal and asthmatic subjects (Curry, 1946;
Casterline and Evans, 1977; Woenne et al, 1978; Nogrady and
Bevan, 1978; Thomson and Kerr, 1980; Eiser et al, 1980, 1981),
indicating that much of the bronchoconstricting effect of
histamine is mediated via the H_1-receptor. This is confirmed
by the protective effect of parenteral H_1-receptor antagonists
on antigen- and exercise-induced asthma (Herxheimer, 1949);
Booij-Noord et al, 1970; Popa, 1980; Zielinski and Chodosowski,
1977; Hartley and Nogrady, 1980; Eiser et al, 1981).

There is some controversy regarding the presence and
function of H_2-receptor in human bronchi, Cimetidine, a

specific H_2-receptor antagonist, given orally intravenously and
by inhalation did not affect bronchial tone in either normal or
asthmatic subjects (Thomson and Kerr, 1980; Eiser et al, 1980,
1981; Nogrady and Bevan, 1981) and in some studies, oral and
inhaled cimetidine did not modify the bronchial response to
histamine, antigen or exercise (Maconochie et al, 1979; Leopold
et al, 1979; Lowhagen and Lindholm, 1979; Thomson and Kerr,
1980; Nogrady and Bevan, 1981). This would imply that H_2-
receptors were not present in human airways. However, two
other groups have disagreed with these conclusions. Nathan and
colleagues (1979) reported that oral cimetidine slightly
accentuated the response of a group of asthmatics to histamine
and so concluded that H_2-receptors were present in their
airways and mediated bronchodilatation. In contrast, Eiser et
al (1980; 1981) postulated that a small population of
bronchoconstricting H_2-receptors may be present in the airways
of normal and asthmatic subjects, since cimetidine 400 mg i.v.
attenuated the bronchial responses to histamine, antigen and
exercise.

Although the effects found by Nathan and Eiser and their
colleagues were statistically significant, they were relatively
trivial and so if a small population of H_2-receptors are
present in human airways, it is unlikely that they are of any
biological significance. The discrepancies in the results
quoted may be due to methodological differences. The results
from some of the studies may have been somewhat misleading,
since, in some cases, the bronchial inhalation techniques were
poorly standardised, in some only single dose-response
relationships were examined and in some, particularly when oral
or inhaled cimetidine had been given, it is possible that the
local dose of antagonist in the lung was inadequate to block
any H_2-receptors present. In addition, the physiological
measurements used to assess responses to histamine and antigen
reflect changes in airways calibre, which may be influenced not
only by changes in bronchial muscle tone, but also by changes
in bronchial secretions and pulmonary vessel calibre. It is
known that, in man, histamine induces pulmonary vasodilation
(Boe and Simonsson, 1980) and bronchial gland hypersecretion
via the H_2-receptor (Shelhamer et al, 1980), as well as some
pulmonary oedema. Thus, it is possible that the small effects
of cimetidine on FEV_1 and sGaw may have been indirect via these
mechanisms. To summarise the evidence regarding histamine
receptors in human airways, it is clear that histamine produces
most of its effect via bronchoconstricting H_1-receptors in the
tracheobronchial tree. If H_2-receptors are present at all,
their role is insignificant. Further careful, in vitro,
studies may clarify the position regarding H_2-receptors.
Radiogand binding studies have identified histamine H_1-

receptors in the brain, and when a specific H_2-receptor ligand
is developed this technique may be used to identify histamine
receptors in human airways in vitro.

The innervation of the lung varies widely between species.
I should like to conclude with a brief description of the
evidence concerning neural receptors in human lung. Much of
this work is summarised in the reviews by Richardson (1979) and
Nadel (1980). The principle neural pathways to the human lung
are the excitatory, cholinergic nerves and the recently
described, inhibitory non-adrenergic non-cholinergic nerves.
The effects of the cholinergic nervous system are well known.
Stimulation of cholinergic receptors with acetylcholine,
methacholine or carbachol, in vivo or in vitro, induces
bronchoconstriction, which is inhibited by pretreatment with
the specific competitive antagonist, atropine. Dose-related
constriction of segmental pulmonary arteries and secretion from
bronchial glands have been demonstrated in vitro with
acetylcholine, whose effect is inhibited by atropine.

The inhibitory nervous system was described recently by
Richardson and Beland (1976), who reported that electrical
field stimulation of human bronchial muscle in vitro, after
pre-treatment with atropine, produced bronchodilatation. This
effect was prevented by tetrodotoxin, but not affected by α or
-adrenergic antagonists, and so they concluded that this
bronchodilataion was neurally-mediated, but not via either
cholinergic or adrenergic nervous system. The mediator is
unknown but may be a neuropeptide.

Postganglionic sympathetic fibres from the cervical
sympathetic ganglia reach ganglia situated in the bronchial
smooth muscle and around bronchial glands. However, no
adrenergic fibres have been identified in the muscle, except
within the ganglia. This constitutes something of a mystery,
since drugs with actions at the adrenergic receptor have
profound effects on the airways and radioligand binding studies
have identified α- and β-receptors in human bronchial smooth
muscle in vitro (Barnes et al, 1980). β-adrenergic receptors
predominate, particularly β_2-receptors, although α_2-receptors
are represented also. Dr. Skidmore will be dealing with the
effects of the β-adrenergic receptor in detail. Suffice it to
say that β_2-adrenergic stimulants are potent bronchodilators in
normal and asthmatic subjects and β_2-receptor antagonists
produce bronchoconstriction in asthmatic subjects. A small
body of in vivo work relates to the role of α-adrenergic
receptors in bronchial muscle. It is reported that stimulation
of these receptors produces bronchoconstriction in asthmatics
(Snashall et al, 1978), while β-receptor antagonism produces

slight bronchodilatation in asthmatics and some inhibition of both histamine- and antigen-induced asthma (Gaddie et al, 1972; Patel and Kerr, 1975). However, the validity of this data is in some doubt because the α-adrenergic agonists and antagonists used in these studies are known to have other non-specific actions.

In conclusion, the development of specific agonists and antagonists and the evolution of new techniques, such as radioligand binding and autoradiography are beginning to unravel the complicated, inter-relating roles of the various receptor systems in vitro. In the past, a great deal of confusion has been perpetrated by poorly standardised, in vivo studies in man, extrapolation of data from other animal species to man, the use of relatively non-specific agonists and antagonists, and the relatively crude and indirect methods of evaluating responses, particularly in vivo. The development of further specific radioligands may sort out some of this confusion.

REFERENCES

Ash, A. S. F. and Schild, H. O. (1966).
  Receptors mediating some actions of histamine.
  Br. J. Pharmacol., 27, 427-439.

Barnes, P. J., Karliner, J. S. and Dollery, C. T. (1980).
  Human lung adrenoreceptors studied by radioligand binding.
  Clin. Sci., 58, 457-461.

Black, J. W., Duncan, W. A. M., Durant, G. J., Ganellin, C. R.
  and Parsons, M. E. (1972).
  Definition and antagonism of histamine H_2-receptors.
  Nature, London, 236, 385-390.

Boe, J. and Simonsson, B. G. (1980).
  Histamine H_1- and H_2-receptors in human pulmonary arteries.
  Bull. Europ. Physiopathol. Resp. 16, 108-109p.

Booij-Noord, H. O., Orie, N. G. M., Berg, W. and de Vries, K.
  (1970).
  Protection tests on bronchial allergen challenge with disodium cromoglycate and thiazinamium.
  J. Allergy, 46, 1-11.

Casterline, C. L. and Evans, R. (1977).
  Further studies on the mechanism of human histamine-induced asthma.
  J. Allergy Clin. Immunol., 59, 420-424.

Chakrin, L. W. and Krell, R. D. (1980).
Histamine receptors in the respiratory system: a review of current evidence.
H_2-Antagonists in Peptic Ulcer Disease and Progress in Histamine Research. (Ed. Torsoli, A., Lucchelli, P. E., Brimblecombe, R.E.W.: Excerpta Medica). 338-346.

Chand, N. (1980).
Distribution and classification of airway histamine receptors: The physiological significance of histamine H_2-receptors.
Advances in Pharmacology and Chemotherapy. 17, 103-131.

Church, M. K. and Gradidge, C. F. (1979).
Histamine H_1-antagonists and histamine release from human lung in vitro.
Br. J. Pharmacol., 66, 68p.

Curry, J. J. (1946).
The action of histamine on the respiratory tract in normal and asthmatic patients.
J. Clin. Invest. 25, 785-799.

Dunlop, L. S. and Smith, A. P. (1977).
The effect of histamine antagonists on antigen-induced contractions of sensitised human bronchus in vitro.
Br. J. Pharmacol., 59, 475p.

Eiser, N. M., Mills, J., McRae, K. D., Snashall, P. D. and Guz, A. (1980).
Histamine receptors in normal human bronchi.
Clin. Sci., 58, 537-544.

Eiser, N. M., Mills, J., Snashall, P. D. and Guz, A. (1981).
The role of histamine receptors in asthma.
Clin. Sci., 60, 363-370.

Gaddie, J., Legge, J. S., Petrie, G., and Palmer, K. N. V. (1972).
The effect of an alpha-adrenergic receptor blocking drug on histamine sensitivity in bronchial asthma.
Br. J. Dis. Chest., 66, 141-146.

Hartley, J. P. R. and Nogrady, S. G. (1980).
Effect of an inhaled antihistamine on exercise-induced asthma.
Thorax, 35, 675-680.

Herxheimer, H. (1949).
Antihistamines in bronchial asthma.
Brit. Med. J., ii: 901-905.

Kaliner, M. (1978).
Human lung tissue and anaphylaxis. The effect of histamine
on the immunological release of mediators.
Am. Rev. Resp. Dis, 118, 1015-1023.

Leopold, J. D., Hartley, J. P. R. and Smith, A. P. (1979).
Effects of oral H_1- and H_2-receptor antagonists in asthma.
Br. J. Clin. Pharmacol., 8, 249-251.

Lichtenstein, L. M. and Gillespie, E. (1973).
Inhibition of histamine release by histamine controlled by
H_2-receptor.
Nature, 244, 287-288.

Lowhagen, O. and Lindholm, N. B. (1979).
Clinical evaluation of some anti-allergic drugs by the use
of antigen challenge test.
The Mast Cell: its role in Health and Disease.
(Pitman Medical Publishers) (Ed: Pepys, J. and Edwards, A.
M.) 332-336.

Maconochie, J. G., Woodings, E. P. and Richards, D. A. (1979).
Effects of H_1- and H_2-receptor blocking agents on histamine-
induced bronchoconstriction in non-asthmatic subjects.
Br. J. Clin. Pharmacol., 7, 231-236.

Muittari, A. (1968).
The value of the Metacholine test as a diagnostic method in
bronchospastic disorders.
Ann. Med. Int. Fenn, 57, 197-203.

Nadel, J. A. (1980).
Autonomic regulation of airway submucosal gland secretion.
In: Airway Reactivity: mechanisms and clinical relevance.
Ed: Hargreave, F. E. (Astra Pharmaceuticals, Canada Ltd.),
54-58.

Nogrady, S. G. and Bevan, C. (1978).
Inhaled antihistamines, bronchodilatation and effects on
histamine and methacholine-induced bronchoconstriction.
Thorax, 33, 700-704.

Nogrady, S. G. and Bevan, C. (1981).
H_2-receptor blockade and bronchial hyper-reactivity to
histamine in asthma.
Thorax, 36, 268-271.

Patel, K. R. and Kerr, J. W. (1975).
Effect of alpha receptor blocking drug, thymoxamine, on
allergen induced bronchoconstriction in extrinsic asthma.
Clin. Allergy, 5, 311-316.

Popa, V. T. (1977).
Bronchodilating activity of an H_1-blocker, Chlorpheniramine.
J. Allergy Clin. Immunol, 59, 54-63.

Popa, V. T. (1980).
Effect of an H_1-blocker, Chlorpheniramine, on inhalation
tests with histamine and allergen in allergic asthma.
Chest, 78, 442-451.

Richardson, J. and Beland, J. (1976).
Noadrenergic inhibitory nervous system in human airways.
J. Appl. Physiol., 41, 764-771.

Richardson, J. (1979).
Nerve supply to the lungs.
Am. Rev. Resp. Dis., 119, 785-802.

Shelhamer, J. H., Marom, Z. and Kaliner, M. (1980).
Immunologic and neuropharmacologic stimulation of mucous
glycoprotein release from human airways in vitro.
J. Clin. Invest., 66, 1400-1408.

Snashall, P. D., Boother, F. A. and Sterling, G. M. (1978).
The effect of adrenoreceptor stimulation on the airways of
normal and asthmatic man.
Clin. Sci., 54, 283-289.

Thomson, N. C. and Kerr, J. W. (1980).
Effect of inhaled H_1 and H_2-receptor antagonists in normal
and asthmatic subjects.
Thorax, 35, 428-434.

Woenne, R., Kattan, M. and Levison, H. (1979).
Sodium cromoglycate-induced changes in the dose-response
curve of inhaled methacholine and histamine in asthmatic
children.
Am. Rev. Respir. Dis., 119, 927-932.

Zielinski, J. and Chodosowski, E. (1977).
    Exercise-induced bronchoconstriction in patients with
    bronchial asthma, its prevention with an antihistaminic
    agent.
    Respiration, 34, 31-35.

DISCUSSION

SPEAKER: N. EISER   **CHAIRMAN: G. CUMMING**

RICHARDSON: I noticed that you hardly mentioned the possibility that histamine might produce some of its effect via reflex mechanisms. Is this because there is no evidence on this subject or because you think that histamine does not produce reflex bronchoconstriction?

EISER: There has been a great deal of work on this subject which has produced conflicting results. Some workers have inhibited the bronchial response to histamine with atropine, but it was not clear whether this was through its specific anticholinergic properties or whether it was due to the non-specific effect of the initial bronchodilatation produced by the atropine. Recently, I tried to resolve this issue by comparing the effects of inhaled placebo and atropine (1.5 mg) on histamine-induced bronchoconstriction in a group of normal and asthmatic subjects. The atropine increased baseline sGaw in normal and non-atopic asthmatic subjects by 45 - 50% and in atopic asthmatics by 100%. Initial baseline sGaw values varied considerably within subjects. By repeating the studies on a number of occasions, I was able to compare the effects of histamine when sGaw, following premedication, was similar (less than 20% variation) within subjects. In these circumstances, atropine greatly attenuated the mean histamine response of both normal and asthmatic groups. The mean antigen response of the atopic asthmatics was also significantly inhibited by atropine. However, in some subjects atropine produced no inhibition of either histamine- or antigen-induced asthma, whereas in others, the effect was dramatic. I think, therefore, that the importance of the parasympathetic nervous system in mediating induced asthma varies widely between individuals.

LEE: In your introduction you reminded us that stimulation of the H1-receptor produced not only bronchoconstriction, but also pulmonary vasodilatation and possibly pulmonary oedema. Is there any evidence that an increase in interstitial fluid in the small airways plays a part in airways narrowing?

EISER: In various animals inhaled histamine can produce pulmonary oedema, but there is little data in man. It is known that interstitial fluid increases airways resistance, but I do not know of any work in which the oedema was quantified and correlated with small airways function.

CORRIN: When you spoke of the non-adrenergic, non-cholinergic nerves, I thought that you said that they had not been demonstrated structurally.

EISER: I think that is correct.

CORRIN: The Hammersmith workers have demonstrated VIP in the upper airways (nose, pharynx and larynx) and bombesin in the lower airways. In the airways they have also demonstrated by immunocytochemical techniques the presence of nerve fibres containing substance P, supplying bronchial epithelium and muscle and small blood vessels.

EISER: Thank you very much. I knew that the Hammersmith workers had demonstrated VIP and bombesin in human airways, because I have collaborated in some of the work, but I was unaware that they had demonstrated the nerves.

CHAIRMAN: Are you referring to the purinergic system Brian?

JEFFERY: It is the peptidergic system which has been demonstrated and discussed rather than the purinergic system. It is interesting to note that substance P and VIP were found in the nerves outside or beneath the epithelium, particularly surrounding the glands, whereas bombesin was found within epithelial cells but not in nerves. This distribution has been demonstrated in several species, including cat and man.

DENISON: Noemi, towards the end of your formal talk you said that the lung had two important nervous systems, the cholinergic and non-adrenergic, non-cholinergic systems, and that stimulation of the cholinergic system produced, not only marked bronchoconstriction, but also pulmonary vasoconstriction. Is there any evidence that the vasoconstriction is secondary to the bronchoconstriction?

EISER: This is work done on segmental pulmonary arteries in vitro and so there were no bronchi present. Presumably therefore, this was a primary effect on the pulmonary vessels, which was inhibited by a competitive antagonist, atropine.

DENISON: What about the existing tone of the vessels?

EISER: I am really not competant to answer this, since I have not done any in vitro work of this type. The data to which I referred, comes from Simonsson's group in Lund.

COBO: How does atropine block the histamine response?

EISER: Atropine is a competitive antagonist of the parasympathetic nervous system. It is presumed that histamine must be acting indirectly, via the vagus nerves, as well as directly on bronchial smooth muscle. In my own experiments it appeared that the contribution of the vagus varied between subjects, both the normal and asthmatic subjects.

BRONCHODILATORS

I. F. Skidmore

Glaxo Group Research Limited

Ware, Hertfordshire, SG12 ODJ

The Pathobiology of Asthma (1)

Asthma can be defined as a reversible increase in airways resistance associated with hyper-reactivity to irritant stimuli and with mechanical obstruction of flow in these airways. Although hyper-reactivity is an important determinant of the clinical response, little is known about its pathogenesis and treatment is directed mainly against obstruction. Physiological and pathological investigation of the disease shows that there are three main causes of obstruction; bronchoconstriction, inflammation and mucus production. In mild asthma bronchoconstriction predominates but as the disease increases in severity, inflammation and mucus become increasingly important until in status asthmaticus a combination of all three causes almost total obstruction.

Current views of the pathobiology of allergic asthma ascribe a central role to the mast cells of the airways (2).

These cells are sensitised to antigen by specific IgE carried on the cell membrane. Cross-linking of these antibody molecules initiates a series of biochemical reactions leading to the release of mediators which induce both bronchoconstriction and inflammation. Histamine and slow reacting substance of anaphylaxis (SRS-A, leukotrienes C_4 and D_4) are the best characterised of these mediators but prostaglandins and chemotactic factors for neutrophils and eosinophils may also contribute to the development of inflammation. Both histamine and SRS-A contract bronchial smooth muscle directly and, in addition, histamine activates

31

irritant receptors in the pulmonary epithelium, triggering a
vagal reflex and cholinergic bronchoconstriction.

Classes of Drugs

In theory bronchoconstriction can be treated either by
pharmacological antagonism of the agent causing
bronchoconstriction using specific antagonists (H_1 antagonists,
SRS-A antagonists, anticholinergic drugs) or by direct
bronchodilatation using drugs that relax smooth muscle
directly. In practice the only useful pharmacological
antagonists are the anticholinergic atropine derivatives which
inhibit vagally mediated bronchoconstriction. With the
exception of these drugs effective reversal of bronchospasm is
only achieved with drugs that relax smooth muscle directly.
Only two classes of drugs of this type are in common use today,
the β_2 selective adrenergic stimulants and the methylxanthines.

Beta$_2$ adrenoceptor stimulants (3)

(a)   Development

The development of β_2 stimulant bronchodilators was a
consequence of the recognition of the value and the limitations
of adrenaline and isoprenaline as bronchodilators. The work of
Ahlquist (4) and of Lands and his colleagues (5) allowed the
rationalisation of the many pharmacological actions of these
drugs in terms of their interaction with three distinct
receptors, alpha, beta$_1$ and beta$_2$ and signalled the advantages
that a drug acting selectively at β_2 receptors would have as a
bronchodilator with minimal cardiovascular side effects. At
the same time understanding of the metabolism of these drugs
emphasised the importance of the catechol nucleus both for
uptake and for metabolism in the gut, liver and lungs.
Chemical modification of these catecholamines was therefore
undertaken, the outcome of which was a range of bronchodilators, most of which are selective for β_2 receptors and
which as they are resistant to metabolism are active by mouth
as well as by inhalation and are of relatively long duration of
action (Table 1).

(b)   Mechanism of action (6)

The β_2 receptors on the external surface of the plasma
membrane of the smooth muscle cell are linked to the enzyme
adenylate cyclase situated on the internal surface of the
membrane. Occupation of the receptor by a β_2 selective agonist
activates adenylate cyclase, catalysing the conversion of ATP

to cyclic AMP. The elevation of cyclic AMP within the smooth
muscle cell inhibits the contraction of smooth muscle fibres as
follows:

The contraction of smooth muscle is controlled by the
interaction of myosin with actin to form the contractile actin-
myosin complex. For this to occur the light chain component of
myosin must be phosphorylated and this phosphorylation is
catalysed by a calcium-dependent enzyme myosin light chain
kinase. This enzyme exists in both active and inactive form
and conversion of the active form to the inactive form is
catalysed by a protein kinase which phosphorylates light chain
kinase rendering it inactive. This inhibitory protein kinase
is itself activated by cyclic AMP. Thus elevation of
intracellular concentrations of cyclic AMP as a consequence of
β_2 receptor occupation inactivates myosin light chain kinase,
prevents the phosphorylation of myosin and its subsequent
interaction with actin to form the contractile complex. In
this way the contraction of smooth muscle is prevented or
reversed.

(c) Route of Administration

     Most of the β_2 stimulant bronchodilators used in the
treatment of asthma have been formulated both as tablets and as
aerosols or dry powders for inhalation (Table 1). For a
variety of reasons practice varies throughout the world but in
the United Kingdom the inhaled route of administration is
greatly preferred. One immediate advantage is the rapid onset
of bronchodilatation achieved by this route. The other
advantage is the elimination of possible side effects. The
target organ is the lung and within the lung the smooth muscle
and mast cells of the airways are the target cells. It is
therefore entirely rational to administer the drug as directly
as possible to these cells. As inhaled drug reaches these
cells without being exposed to metabolism in gut and liver and
does not have to be distributed throughout the body in order to
reach the lungs the inhaled dose is ten to twenty times less
than the oral dose. The concentration of drug in the systemic
circulation following inhalation is therefore much lower than
that following an equipotent oral dose and the possibility of
side effects is reduced accordingly.

(d) Frequency of administration

     In mild episodic asthma β_2 stimulant bronchodilators,
preferably by inhalation, can be prescribed to be taken as
required, one or two puffs producing bronchodilatation of rapid
onset and sufficient duration to outlast the attack. In

Table 1

Sympathomimetic Bronchodilators in Common Use

	Inhaled Dose (μg)	Oral dose (mg)	β₂ Selectivity	Duration (hrs)
Isoprenaline	80-400	NA	No	2
Fenoterol	200-400	NA	Yes	4-8
Orciprenaline	750-1500	20	No	3-4
Terbutaline	250-500	5	Yes	4-6
Salbutamol	100-200	2-4	Yes	4-6
Clenbuterol	NA	10-20	Yes	> 12

NA   not available

Table 2

Activities of β Adrenoceptor Agonists in the Anaesthetised
Guina-Pig. Drugs Adnimistered Intravenously

	Decrease tracheal pressure	Equipotent Dose Decrease soleus muscle tension	Increase heart rate
(-)Isoprenaline	1	1	1
Orciprenaline	102	107	304
Fenoterol	8	6	159
Terbutaline	29	34	257*
Salbutamol	10	9	74*

* Partial agonist compared to (-)Isoprenaline

From G.P. Levy + G.H. Apperley, (1978). (Reference 10).

chronic or recurrent asthma they can be prescribed in the same
way but modern practice is to prescribe these drugs to be taken
regularly three or four times a day in a prophylactic regimen.
This has the advantage of providing continuous protection
against attacks that may occur throughout the day.   In
addition, however, there is evidence that this regimen reduces
the hyper-reactivity of the airways and lowers the incidence
and severity of nocturnal asthma (7).   The pharmacological
basis for this reduction in hyper-reactivity is not clear but
we believe that it reflects the ability of β_2 selective drugs
such as salbutamol not only to relax smooth muscle but also to
inhibit the release of histamine and SRS-A from the mast cells
of the lung by interacting with β_2 receptors on these cells
(8).   The effect of this will be to reduce not only the
bronchoconstriction caused by these mediators but to reduce the
inflammation that they cause. Airways hyper-reactivity appears
to be related to the severity of pulmonary inflammation (9).

(e)   Side Effects and Toxicity

The β_2 selective bronchodilators are without significant
side effects when taken by inhalation at the recommended dose.
When taken by mouth both skeletal muscle tremor and tachycardia
are seen.   Study of a wide range of β_2 specific bronchodilators
has failed to show a separation between bronchodilator and
tremorigenic activity (10) (Table 2), indicating that tremor is
mediated through β_2 receptors.   The origin of the tachycardia
is more complex although the greater part of the response is
again due to activity at β_2 receptors.   Peripheral
vasodilatation, mediated by β_2 receptors, leads to reflex
tachycardia and in addition there are a minority of β_2
receptors in the heart through which direct chronotropic
effects of β_2 selective bronchodilators may be exerted (10).
Cardiovascular responses even to high oral doses of salbutamol
are remarkably mild (3) and deliberate attempts at self
poisoning with this drug have been singularly unsuccessful
(11).   The increase in deaths from asthma seen in the United
Kingdom in the ninteen sixties before the introduction of β_2
selective bronchodilators was not repeated following their
introduction.   Deaths from asthma have remained essentially
constant in the thirteen years that these drugs have been
available while the annual prescription rate has increased from
500,000 units in 1969 to well in excess of 5 million in 1980.
Reports of deaths following overdosage with β_2 stimulant
bronchodilators in all probability reflect not the "toxicity"
of the drugs but a grim reminder that acute severe asthma can
be a fatal disease if not treated correctly.   It is of
paramount importance that the physician educates the patient to
recognise that increasing use of a bronchodilator without

corresponding relief is a sign of deteriorating asthma requiring additional medication.

(f)   Tolerance

It has been suggested that pharmacological tolerance to β stimulant bronchodilators may develop following repeated high doses of these drugs and that this may have been a contributory factor in the increase in deaths from asthma seen in the mid sixties. Normal subjects given high doses of salbutamol developed tolerance to the metabolic and bronchodilator effects of the drug. However, when these experiments were repeated in asthmatic subjects although tolerance to the metabolic effects (increases in plasma glucose, pyruvate and cyclic AMP) was observed there was no change in the airway response to salbutamol (12). Thus the airways of asthmatics appear to be selectively protected from developing resistance and differ from the airways of normal subjects. The physiological and biochemical basis for this difference is unknown.

Methylxanthines (13)

(a)   Development

Methylxanthine bronchodilators have been used in the treatment of asthmatic bronchoconstriction for many years. By far the most common of these drugs is theophylline (1-3-dimethylxanthine) which has been formulated in many different ways, as a soluble complex with ethylenediamine, as a soluble salt with choline, as microcrystalline anhydrous theophylline and as a series of controlled release preparations designed to maximise bioavailability and provide therapeutic concentrations in the blood for up to twelve hours. The pharmaceutical ingenuity that has gone into these varied formulations has originated mainly in the United States. It reflects the dependence of American physicians on theophylline as the β_2 stimulant bronchodilators available in Europe for the last thirteen years are only now becoming freely available in the United States. It will be interesting to follow the relative fortunes of these very different bronchodilators in the American market in the next few years.

(b)   Mechanism of action

Theophylline and other methylxanthines inhibit the enzyme cyclic AMP phosphodiesterase which inactivates cyclic AMP by hydrolysing it to 5'AMP. It has been assumed for many years that this activity is the basis of the bronchodilator action of

these drugs. Adenylate cyclase has a low basal activity that
is expressed even in the absence of activation by hormones or
neurotransmitters and so all cells contain low concentrations
of cyclic AMP. These concentrations reflect the balance
between cyclase activity and phosphodiesterase activity.
Inhibition of phosphodiesterase will therefore increase the
concentration of cyclic AMP in the cell by inhibiting breakdown
without affecting synthesis. Elevation of cyclic AMP by this
mechanism will cause relaxation of smooth muscle by the same
mechanism as will elevation by stimulation of adenylate
cyclase.

In recent years, however, this view of the mechanism of
action of theophylline has been challenged (14). It has been
pointed out that, at therapeutic concentrations (10 - 20
ug/ml), the activity of theophylline as a phosphodiesterase
inhibitor is very low and little elevation of cyclic AMP can be
expected under these conditions. Alternative suggestions as to
the mechanism of action of theophylline have been made. The
receptor-adenylate cyclase system of the plasma membrane
depends for its action not only on occupation of the receptor
but on modulation of the signal between occupied receptor and
cyclase. Adenosine appears to be an important inhibitory
modulator of cyclase activity, controlling the activity of the
enzyme in many cell types. Theophylline, in addition to being
a phosphodiesterase inhibitor is a competitive antagonist of
the action of adenosine in this inhibitory role and is of such
potency that, at therapeutic concentrations, it will have
considerable activity as an antagonist. It has therefore been
proposed that theophylline acts as a bronchodilator by reducing
endogenous inhibition by adenosine of adenylate cyclase in
smooth muscle cells (14). The consequence of this would be an
increase in intracellular cyclic AMP and reversal or prevention
of the contractile event. A third proposal is that
theophylline prevents or reverses contraction by altering the
availability of calcium ions for the contractile process (15).

(c) Administration

The relatively low potency of theophylline and its
derivatives as bronchodilators precludes their formulation in a
pressurised aerosols although dry powder formulation may be
possible. Theophylline preparations currently available are
primarily for oral use but the soluble salts are also used for
intravenous administration.

Theophylline is a drug with a very low therapeutic index
and unpleasant, potentially dangerous, side effects. Even at
therapeutic doses significant side effects are seen in a

minority of patients. The development of rapid and accurate
methods for measuring the concentration of theophylline in the
plasma and the formulation of preparations with reliable
release rates and high bioavailability has made the physicians
job much easier but the variability in patient response both in
terms of bronchodilatation and side effects even within the
accepted therapeutic concentration range (10 - 20 ug/ml plasma)
makes measurement of plasma concentrations almost mandatory if
effective treatment is to be achieved. In general oral dosage
should not exceed 15 mg/kg per day in divided doses.
Intravenous aminophylline is still frequently used in the
treatment of acute severe asthma. It must not be given rapidly
as severe cardiovascular side effects can result. Normal
practice is to give a loading infusion of 5 - 6 mg/kg body
weight over 20 minutes to achieve a concentration in the plasma
of between 10 and 20 ug/ml rapidly and safely and to follow
this with a steady infusion of 0.7 - 0.9 mg/kg per hour to
maintain this concentration. If the patient is currently
taking oral preparations containing theophylline intravenous
doses must be lowered. Monitoring of plasma concentration is
of great importance during intravenous administration.

(d)   Side effects and toxicity

     Theophylline and its derivatives produce a wide range of
potentially dangerous side effects and there is considerable
danger of self poisoning with these drugs (11). Side effects
are dose related and can restrict the degree of
bronchodilatation achieved with these drugs alone. A
proportion of subjects experience gastric disturbances, nausea,
vomiting, anxiety and headache with therapeutic doses and this
can severely limit patient compliance. The introduction of
controlled release preparations has reduced the incidence of
side effects in this range but if the concentration of
theophylline in the plasma significantly exceeds 20 ug/ml there
is danger of cardiac arrhythmias leading to cardiac arrest,
cerebral seizures and respiratory arrest. These side effects
may be due to one or more of the proposed biochemical
mechanisms by which these drugs work but at this time there is
no definitive evidence to incriminate phosphodiesterase
inhibition, adenosine antagonism or effects on calcium
distribution (16) although it has been suggested that adenosine
antagonism is responsible for side effects in the central
nervous system (17).

     The potential toxicity of theophylline is of particular
importance in patients with impaired liver function. Reduced
hepatic metabolism may result in toxic concentrations in the
plasma following normal therapeutic doses. Rapid intravenous

injection, leading to elevated concentrations of drug in the blood, must be avoided and it is emphasised that intravenous dosing must take account of any current oral therapy.

(d)   Drug combinations

Maximum bronchodilatation is rarely achieved with oral methylxanthines alone and frequently these drugs are given in combination with β-stimulant bronchodilators. Several combination preparations with ephedrine and barbiturates are available but quantitative evidence that these perform better than theophylline alone is lacking. More frequently theophylline preparations are coprescribed with either oral or inhaled β_2 stimulant bronchodilators. If theophylline is acting as a phosphodiesterase inhibitor combined therapy would be expected to lead to a synergistic effect as the increase in intracellular cyclic AMP produced by both stimulating synthesis and inhibiting breakdown of cyclic AMP would be greater than the sum of the two effects separately. Clinical studies designed to investigate the possibility of synergy between the two classes of drug almost unanimously demonstrate that additive and not synergistic bronchodilator responses are obtained (18 - 20). This failure to demonstrate synergy does not, however, invalidate the combined use of the two drugs in those patients in whom, for various reasons, adequate bronchodilatation is not achieved with a single drug.

SUMMARY

True bronchodilatation, as opposed to antagonism of bronchoconstriction is achieved with two classes of drug, the β-adrenergic stimulants and the methylxanthines. Of the β-stimulants those which act selectively at β_2 adrenoceptors have the advantage of producing effective bronchodilatation with minimal cardiovascular side effects. They act by increasing the concentration of cyclic AMP in smooth muscle thus inhibiting the contractile process. Administration of these drugs by inhalation from pressurized aerosols, dry powder formulations or nebulised solutions virtually eliminates side effects. When taken by mouth the major side effect is skeletal muscle tremor, which, as it is mediated through β_2 receptors cannot be eliminated. Tachycardia is a side effect consequent on peripheral vasodilatation and perhaps on activation of a minority of β_2 receptors mediating direct chronotropic effects in the heart. Tolerance to bronchodilator action does not appear to develop in asthmatics. These drugs can be taken on demand in mild asthmatics but in chronic or recurrent asthma they are increasingly prescribed in a prophylactic regimen. This, in addition to providing continuous bronchodilator action, also reduces airways hyper-reactivity improving night-time control. Reduction in hyper-reactivity may be related to the ability of these drugs to inhibit the release of spasmogens and inflammagens from the mast cell.

Methylxanthines, primarily theophylline and its soluble salts are taken by mouth or, in acute severe asthma, given by intravenous infusion. Although these drugs inhibit the inactivation of cyclic AMP by phosphodiesterase and thus raise intracellular concentrations of cyclic AMP alternative mechanisms of action have been proposed. These drugs have potentially serious side effects which limit patient compliance and the degree of bronchodilatation that may be achieved. Good therapy with these drugs depends on adjusting the dose carefully to individual responses and ideally plasma concentrations of drug should be monitored. Combined therapy with theophylline and β_2 stimulant bronchodilators produces additive rather than synergistic effects.

On balance it appears that β_2 stimulants given by inhalation combine maximal effectiveness with minimal side effect liability and should be considered as the drugs of choice. It will be interesting to monitor the penetration of these drugs into the American market which up to now has relied mainly on the methylxanthines.

REFERENCES

1.   Asthma (1977) (eds. T. J. H. Clark & S. Godfrey) Chapman
         and Hall Limited, London.

2.   Skidmore, I.F. (1982).
         Allergic Asthma and Rhinitis: the relationship
         between pathobiology and treatment.
         Trends Pharmacol. Sci., 3, 66 - 70.

3.   Brittain, R. T., Dean, C. M., & Jack, D. (1981).
         Sympathomimetic Bronchodilator Drugs, in Respiratory
         Pharmacology pp. 613 - 655, (International
         Encyclopedia of Pharmacology and Therapeutics section
         104. ed. J. G. Widdicombe). Pergamon Press, Oxford.

4.   Ahlquist, R. P. (1948).
         A study of the adrenotropic receptors.
         Amer. J. Physiol., 153, 586 - 600.

5. Lands, A. M., Arnold, A., McAuliff, J. P., Luduena, F.P. & Brown, T. G. (1968).
Differentiation of receptor systems activated by sympathomimetic amines.
Nature (Lond). 214, 597 – 598.

6. Stull, J. T., Blumenthal, D. K., & Cooke, R. (1980).
Regulation of contraction by myosin phosphorylation; a comparison between smooth and skeletal muscles.
Biochem. Pharmac., 29, 2537 – 2543.

7. Clark T. J. H. (1980).
Diurnal variation in airway obstruction: clinical significance.
IM Internal Medicine for the specialist, 1, 62 – 68.

8. Butchers, P. R., Skidmore, I. F., Vardey, C. J. & Wheeldon, A. (1980).
Characterisation of the receptor mediating the antianaphylactic effects of adrenoceptor agonists in human lung in vitro.
Br. J. Pharmac., 71, 663 – 667.

9. McFadden Jnr., E. R., Soter, N. A. & Ingram Jnr., K. H. (1980).
Magnitude and site of airway response to exercise in asthmatics in relation to histamine levels.
J. Allergy Clin. Immunol., 66, 472 – 477.

10. Levy, G. P., & Apperley, G. H. (1978).
Recent advances in the pharmacological subclassification of adrenoceptors, in Recent Advances in the Pharmacology of Adrenoceptors. pp. 201 – 208. (eds. E. Szabadi, C. M. Bradshaw & P. Bevan) Elsevier/North Holland.

11. Prior, J. G., Cochrane, G. M., Raper, S. M., Ali, C. & Volans, G. N. (1981).
Self-poisoning with oral salbutamol.
Br. Med. J., 282, 1932.

12. Harvey, J. E., Baldwin, C. J., Wood, P. J., Alberti, K. G. M. M. & Tattersfield, A. (1981).
Airway and metabolic responsiveness to intravenous salbutamol in asthma: effect of regular inhaled salbutamol.
Clin. Sci., 60, 579 – 585.

13. Long Term Theophylline Therapy, (1980).
     Europ. J. Resp. Dis., 61, Supplement 109.

14. Fredholm, B. B. (1980).
     Are methylxanthine effects due to antagonism of
     endogenous adenosine?
     Trends Pharm. Sci., 1, 129 - 132.

15. Kolbeck, R. C., Speir, W. A., Carrier, G. O. & Bransome
     Jnr. E. D. (1979).
     Apparent irrelevance of cyclic nucleotides to the
     relaxation of tracheal smooth muscle induced by
     theophylline.
     Lung, 156, 173 - 183.

16. Cardinali, D. P. (1980).
     Methylxanthines: possible mechanisms of action in
     the brain.
     Trends Pharmac. Sci., 1, 405 - 407.

17. Persson, C. G. A., Erjefalt, I., & Karlsson, J. A. (1981).
     Adenosine antagonism, a less desirable characteristic
     of Xanthine asthma drugs.
     Acta Pharmac. Tox., 48, 317 - 320.

18. Svedmyr, K. (1981).
     $_2$ adrenoceptor stimulants and theophylline in asthma
     therapy.
     Europ. J. Resp. Dis. 62. Supplement 116.

19. Marlin, G. E., Hartnett, B. J. S., Berend, N., & Hacket,
     N. B. (1978).
     Assessment of combined oral theophylline and inhaled
     adrenoceptor agonist bronchodilator therapy.
     Br. J. Clin. Pharmac., 5, 45 - 50.

20. Clark, C. J., & Boyd, G. (1980).
     Combination of aminophylline (Phyllocontin Continus
     Tablets) and salbutamol in the management of chronic
     obstructive airways disease.
     Br. J. Clin. Pharmac., 9, 359 - 364.

DISCUSSION

SPEAKER: SKIDMORE   **CHAIRMAN: CUMMING**

QUESTIONER: Recalling the slide that you showed of the mortality from asthma over the period 1955-1980, my recollection is that the mortality rate reached a maximum in the mid sixties, before the introduction of selective B_2 stimulants, but that following the introduction of these drugs mortality remained constant even though prescription levels have increased twentyfold from the year of introduction. I think you ascribed the increase and subsequent decline in deaths to variation in exposure but how would you comment on the constant death rate since the introduction of B_2 stimulants, it could be argued that these drugs have not influenced the death rate from asthma at all.

SKIDMORE: To take your first point, some workers have ascribed the increase in deaths in the mid sixties to an increase in aerobiological challenge putting a higher proportion of severe asthmatics at risk. Dealing with your second point, we must not forget that for a proportion of patients of different ages asthma is always a life-threatening disease. For these patients in particular, bronchodilators alone are inadequate and steroid therapy is also essential. The thousand or so deaths a year from asthma contains a considerable number of patients with chronic severe asthma who succumb to an acute crisis in the disease. A proportion of the deaths are of patients who do not recognise that their asthma is deteriorating and therefore do not revisit their physician to tell him this. The effectiveness of bronchodilators in most asthmatics can, in those whose asthma is getting worse, be a disadvantage as the temptation is to go on increasing the dose rather than seeking advice. To some extent the patient is to blame but it is primarily the doctor's responsibility to educate the patient to notice what is happening and to come back for examination straight away.

CHAIRMAN: One should remember that the peak in asthma deaths was not distributed uniformly between age groups, in the age group 10-14 deaths increased by a factor of 5 while in the 31-40 year group deaths only increased by 20%, so that whatever this factor was, it was affecting the age groups differently.

SKIDMORE: Yes, I think we will go on arguing about the origins of that peak for a long time.

LEE: In your introduction you showed a slide of bronchoconstriction superimposed on inflammation, went on to discuss the role of mast cells in histamine release and to describe the action of B_2 stimulants on histamine release. Williams, from the Pharmacology Department of the Royal College of Surgeons, showed a potentiation of the effects of histamine on skin inflammation by the prostaglandins. The question I would like to ask you is: what are the experimental pharmacologists doing about combination therapy between the B_2 agonists and the appropriate antiprostaglandins.

SKIDMORE: I think there is a problem here, if you are referring to prostaglandin antagonists as antiprostaglandins, there are no antagonists for the particular prostaglandin receptor through which vasodilatation is mediated As far as inhibitors of prostaglandin synthesis are concerned I am sure that we are all aware that in a proportion of asthmatics these are very dangerous drugs indeed. However, in asthmatics showing a late response to antigen this late response is blocked by indomethacin implying that a cyclo-oxygenase product (prostaglandin or thromboxane) is involved in this bronchodilator-resistant phase of the asthmatic response. Bearing in mind that this phase is also steroid-sensitive I would suggest that it is primarily inflammatory in nature.

LEWIS: You mentioned some work presented by Van As showing that long-term treatment with B_2 agonists improved asthma. I believe that this does not agree with some other work, that you presented later in your talk, by Tattersfield and her group that showed that over a months treatment with B_2 agonists failed to improve asthmatics. I wonder whether you have any comment on the apparent contradiction in these two sets of results.

SKIDMORE: No, that fall in specific airways conductance that
                 you see in Tattersfield's data (reference 12) is
                 inexplicable to me, I don't think the fall is
                 significant but how it relates to the data of Van
                 As I don't know. The therapeutic regimen used was
                 in fact very different, Harvey et al used only
                 aerosol doses of salbutamol whereas in the study
                 reported by Van As both oral and inhaled B
                 angonists were used. Whether or not this could be
                 the basis of the difference I don't know, I would
                 doubt it, but I am not sure how consistent the
                 observation of a fall in conductance following
                 repeated aerosol dosing would be. What we really
                 need is more data.

PRICE: May I come back to that slide which shows a
                 plateau where the number of deaths from asthma
                 remains constant, in spite of the fact that there
                 is a great increase in the number of prescriptions
                 for B_2 agonists. Now there is another possible
                 explanation for that which is that the increasing
                 number of prescriptions does not relate to the
                 number of patients. Is the increase in
                 prescriptions due to more prescriptions for the
                 same number of patients and are there many
                 patients who are not getting appropriate therapy,
                 in which case the death rate could remain the
                 same.

SKIDMORE: I think there are two points to be made here. The
                 first is that over the past thirteen years there
                 has been a great increase in the number of
                 patients receiving drugs of this type. The second
                 is that one of the reasons for this is that the
                 diagnosis of asthma has improved greatly over the
                 last 20 years so that the death rate represents a
                 much smaller proportion of identified asthmatics
                 than it did in 1962. This increase in identified
                 asthmatics however does not represent an increase
                 in the number of people with the disease but an
                 improvement in diagnosis.

BONSIGNORE: The development of nocturnal asthma sometimes
                 involves extremely complex pathogenetic
                 mechanisms, in some cases related to sleep
                 disturbances. I would like to know whether you
                 have any experience of the prophylactic effect of

B_2 stimulants in relation to the possible pathogenesis of nocturnal asthma. Furthermore, what is your experience of the comparison between delayed-release theophllyine and B_2 stimulants.

SKIDMORE: I can't speak about the pathogenetic mechanisms, I don't think we understand sufficiencly the nature of nocturnal asthma. One comment I can make is to repeat what I said in my talk, that nocturnal asthma appears to be related to the degree of hyper-reactivity which itself can be reduced by good daytime control of asthma using a prophylactic regimen. Most attempts to control nocturnal asthma using controlled release preparations do not appear to have been very successful but recently it has been shown that high dose controlled release aminophylline was quite effective but the doses needed were greater than those used for daytime control (1). There is a feeling that high doses of B_2 stimulants would provide the same relief. The problem is that with oral bronchodilators and theophylline in particular side effects become limiting and we cannot go on increasing the dose in the anticipation of achieving a therapeutic effect.

EISER: I would like to make a couple of comments about several other things you mentioned. One is about nocturnal asthma, I agree with you that no one has managed to produce a good explanation for this but Barnes (2) showed that plasma histamine rises at night and plasma adrenaline is low at night: I think there may be some indication as to pathogenesis from this. In addition some time ago it was shown that people are more responsive to histamine at night (3), so there may be many possible reasons for nocturnal asthma. With regard to long-term salbutamol a recent study showed that bronchial reactivity as judged by histamine dose-response curves, did not alter significantly (4). Finally, in the South African study it wasn't clear whether the peak flow values were a mean of those taken over a month or whether they were just random measurements, it would make a great deal of difference.

SKIDMORE: If I could answer that last point first. I agree that there are a lot of things that are not clear about the South African study, as far as I know we

still do not have the full details of the study so
I cannot say whether the values were means or
random measurements.

DENISON: A small point about the Southampton study, I
believe the first author was Harvey, in which you
showed the dose-response curve for salbutamol
assessed by measuring specific conductance. The
results all appear to be within the normal range,
the small differences before and after treatment
were trivial but I wondered whether as all the
values were in the normal range the whole study
couldn't be considered trivial.

SKIDMORE: I don't think I can answer that question but is
there anyone in the audience who would like to
make a comment about the normality or otherwise of
airway function in the Southampton asthmatics.

LEWIS: Being a Southampton asthmatic myself I comment
that the degree of dilatation seen in these
studies is significant, represents the maximum
amount of dilatation that we would expect and is
by no means trivial. I think that one of the
problems with long-term studies is that more is
changing than just the administration of drug,
over the course of say a month or six weeks the
season may alter, the subject may enter a pollen
season or it might start raining, which it does
quite a lot in England, and there may be other
reasons for the small insignificant fall in
airways conductance in this study.

SKIDMORE: If I could add a comment, perhaps I didn't make it
clear that all other parameters measured that
related to possible desensitisation were changed
by drug treatment, metabolic responses, cyclic AMP
responses, the one thing that was not changed
significantly was the response of the airways.

CHAIRMAN: I wonder if I could be forgiven a short comment on
that question. Asthma is defined, as was clearly
done on the first slide, in terms of the
limitation of airflow, which was called on the
slide obstruction and indeed there may be
obstruction or there may not. Other functional
defects in asthma, for instance the ventilatory
abnormality, are measured during quiet breathing
and this raises the question: is there flow

limitation in quiet breathing in asthmatics. The answer seems to be 'no there is not! Thus although the larger airways are important and B_2 stimulants affect these, there is another component in the disease. This is occlusion of small airways, which is responsible for the ventilatory defect and is insensitive to B_2 stimulants. It may, however, be sensitive to steroids and this may be the effective role of steroids. The common clinical experience that if you give an asthmatic steroids the FEV_1 does not increase but the patient feels better was thought to be due to psychological effects but I believe it is due to the relief of occlusion of the small airways. This may give a little background to this conflict of evidence between measurements of airflow resistance and the dose of B_2 stimulants.

JENNETT: I wonder whether the action of B_2 stimulants is affected at all by the pH and pO_2 of the lung, are there any studies in vitro that might be relevant to the situation.

SKIDMORE: I would not be surprised if responses varied with pH and pO_2, studies have shown that even changes in temperature affect the sensitivity of receptors for catecholamines and some workers have ascribed exercise-induced bronchoconstriction to a desensitisation of B receptors caused by cooling of the airways.

EISER: May I just make one comment about the site of action of salbutamol. There was a study comparing the effects of salbutamol and atropine as bronchodilators, looking at flow volume curves, and the conclusion was that atropine appears to be dilating more centrally and salbutamol more peripherally. Could I also ask about the use of high dose salbutamol, we have a number of chronic asthmatics and patients with chronic reversible airway obstruction which appears to be reversible by very large doses of salbutamol given by nebuliser. Have you any comment on this?

SKIDMORE: Do you know how much drug the patient is getting?

EISER: Not exactly but the patient inhales about 10% of what is nebulised and this is a much greater amount than is given in a pressurised aerosol.

SKIDMORE: I think that if you are getting relief from the
               drug then it is perfectly acceptable to do this
               because the evidence that tolerance develops is
               not good. Why they should need such a high dose,
               however, is a different matter. Do they respond
               adequately to intravenous salbutamol?

EISER: We haven't given them intravenous salbutamol but
               we have a number on nebulisers at home having
               found that they responded well in hospital to a
               large dose.

REFERENCES

1. Barnes, P. J., Greening, A.P., Neville, L., Timmers, J, &
   Poole, G. W. (1982).
   Single dose slow-release aminophylline at night prevents
   nocturnal asthma. Lancet 1: 229-301.

2. Barnes, P., Fitzgerald, G., Brown, M. & Dollery, C. (1980).
   Nocturnal asthma; changes in circulating epinephrine,
   histamine and cortisol.
   New Eng. J. Med., 303, 263-267.

3. DeVries, K., Goei, J. T., Booij-Noord H, & Orie, N.G.M.
   (1962).
   Changes during 24 hours of the lung function and histamine
   hyperactivity of the bronchial tree in asthmatic and
   bronchitis patients.
   Int. Arch. Allergy, 20, 93-101.

4. Peel, E. T. & Gibson, G. J. (1980),
   Effects of long-term inhaled salbutamol therapy on the
   provocation of asthma by histamine.
   Am. Rev. Resp. Dis., 121, 973-977.

HORMONES AND THE LUNG

B. Corrin

Cardiothoracic Institute

Brompton Hospital, London

Hormonal control of lung growth

Liggins reported in 1969 that premature lambs survived better if they had been treated in utero with glucocorticoids and suggested that glucocorticoids promoted foetal lung maturation and hence surfactant production. Subsequent studies have shown that glucocorticoid treatment of the foetus accelerates structural development of the lungs. Glucocorticoid receptors have been identified in higher concentration in the lung than any other foetal tissue. Using isolated cells, glucocorticoid receptors have been identified in the nuclei of lung fibroblasts and type II cells, but the concentration is no higher than in whole lung In keeping with this, the glucocorticoid-enhanced structural maturation of the lung is generalised, suggesting that all cell types are affected.

Liggins subsequently conducted a clinical trial of pre-natal betamethasone treatment for the prevention of the infantile respiratory distress syndrome and reported increased survival of premature infants exposed to maternal steroid therapy and a decreased incidence of the respiratory distress syndrome. Subsequent trials have confirmed this and such treatment is recommended to consenting women, 26-34 weeks' pregnant, threatened by premature labour, in whom delivery can be delayed for at least 24 hours (Ballard and Ballard 1979). The injected dose of betamethasone (6 mg, repeated 24 hours later if necessary) results in a cortisol-equivalent foetal blood level which is in the physiological response to stress range. Pharmacological levels are not reached. Stressful maternal situations such as heroin addiction and hypertension may protect against infantile respiratory distress by increasing glucocorticoid levels and hence promoting maturation of the lung. Experimental manipulations such as in utero decapitation delay the maturation of the foetal lung, probably by reducing foetal ACTH levels.

Natural sources of cortisol available to the foetal lung include the maternal and foetal adrenals, the amniotic membranes and the foetal lung itself. The human placenta permits the passage of steroids but much of the cortisol is converted to inactive cortisone. However, after 20 weeks' gestation the amniotic membranes switch to producing cortisol from cortisone. There is a similar production of cortisol from cortisone in the lung itself and this also increases towards term.

There is evidence that thyroxine, oestrogens and β-sympathomimetics similarly affect lung growth. Thyroxine has been shown to promote morphological maturation of the lung, bubble stability of lung fluid, and choline incorporation into pulmonary phospholipids. Conversely, foetal thyroidectomy results in reduced lecithin/sphingomyelin ratios in tracheal fluid. That oestrogen receptors exist in the adult lung may be inferred from the observation that pulmonary lymphangioleiomyomatosis is confined to women in the reproductive years and ovariectomy appears to arrest the proliferative process which characterises this disease.

Metabolism of hormones and other circulating substances by the lung

Starling and Verney recognised in 1925 that a vasoactive blood substance was inactivated as it passed through the lesser circulation. This substance was identified as 5-hydroxytryptamine (5HT) by Gaddum et al in 1953. It is now appreciated that many substances, including some drugs and hormones, are modified as they pass through the lung and that the pulmonary endothelium is not just a smooth inert vascular lining. Some substances are merely bound to the endothelium and may be dislodged by other factors. Other substances are taken up by the endothelium and actively metabolised. The metabolism may take place on the surface or within the cytoplasm of the endothelial cell and may result in the substance being activated or inactivated. On the other hand, some chemically related substances pass through the pulmonary circulation unchanged.

Imipramine is a drug for which the lung has a high affinity

but no metabolic capability. Further infusions of imipramine
or of chlorpromazine will displace imipramine previously
accumulated in the lung (Junod 1972).

Substances metabolised upon the surface of the pulmonary
endothelial cell include several important peptides and
nucleotides. Bradykinin is largely metabolised on its passage
through the lung and it is notable that the effector peptidase
is also concerned in the metabolism of angiotensin I. That a
single enzyme inactivates bradykinin and activates angiotensin
represents remarkable economy of function. Since this enzyme
cleaves the carboxyl terminal dipeptide from both bradykinin
and angiotensin I it is preferable to use its biochemical name
dipeptidyl carboxypeptidase (EC3.4.15.1) rather than either of
its trivial names, kininase or angiotensin converting enzyme.
Ryan and colleagues (1976) have localised both dipeptidyl
carboxypeptidase and nucleotidase to the luminal surface of the
alveolar capillary endothelial cell by ultrastructural
immunocytochemistry. The former is distributed uniformly along
the cell surface whilst the latter is located within certain
caveolae.

Substances taken up and metabolised inside the endothelial
cell include 5HT (by monoamine oxidase), nor-adrenaline (by
catechol-o-methyl transferase) and prostaglandins of the E and
F series (by 14-hydroxyprostaglandin dehydrogenase). Junod and
colleagues have localised the uptake of 5HT and nor-adrenaline
within the pulmonary blood vessels by autoradiography; the
former is taken up by endothelial cells throughout the
pulmonary circulation whereas the latter is preferentially
taken up by the endothelium lining small pre and post-capillary
vessels and veins (Nicholas et al, 1974).

Despite the presence within the pulmonary endothelial cell
of peptidases capable of splitting oxytocin, vasopressin and
substance P, these substances are unchanged on passage through
the pulmonary circulation, because there is no uptake.
Similarly histamine and prostaglandin A pass through the lung
unchanged despite the presence of intracellular imidazole-N-
methyl transferase and 14-hydroxyprostaglandin dehydrogenase,
again because of the low affinity the lung has for these
substances (Ferriera et al 1980).

Bioactive substances formed in the lung

Several bioactive substances apart from angiotensin II are
formed within the lung, and some of these reach the systemic
circulation. Most of them however are formed within the
interstitial tissues or the walls of airways and their escape

into the circulation is probably accidental. Substances which
fall into this category include mediators of anaphylaxis, such
as histamine and SRS-A, the latter a leucotriene prostanoid.
Indeed several prostanoids are released into the pulmonary
venous blood on antigenic challenge whilst hypoxia,
hyperventilation or mechanical stimulation of the lungs have a
similar effect.

Whereas most prostanoids are formed in the connective
tissues of the lung, prostacyclin synthetase is found
particularly in relation to the arterial intima. Prostacyclin
is spontaneously released from the lungs and it has been
proposed that this represents active secretion by the
endothelium (Gryglewski et al 1978). Prostacyclin is
vasodilatory and with regard to platelets, anti-aggregatory.
In particular, prostacyclin is antagonistic to thromboxane, the
platelet aggregating factor released from platelets themselves.
Prostacyclin therefore plays an important role in minimising
pulmonary thrombosis. In addition to these important local
effects, it is proposed that prostacyclin released from the
lungs has important effects elsewhere in the body, contributing
to the patency of systemic vessels and protecting against
coronary and cerebral insufficiency. The hypothesis is
advanced that the pulmonary endothelium may be considered to
represent one large endocrine gland generating prostacyclin
(Moncada et al 1978).

Endocrine-type cells in the lung

Feyrter (1953) noted the presence of occasional poorly
staining cells in the bronchial epithelium and included them in
his diffuse paracrine or clear cell system. The
ultrastructural features of these cells were described by
Bensch and colleagues (1965, 1968). They are typically found
near the basement membrane, although they often stretch out a
thin cytoplasmic process between the adjacent epithelial cells
to reach the lumen. Their characteristic feature is the
presence of many cytoplasmic granules of neuro-secretory or
endocrine type consisting of a central electron-dense core,
separated from an outer investing membrane by a clear halo,
altogether measuring about 140 nm diameter. Using silver
stains Tateishi (1973) studied the distribution of these cells.
He described small numbers of argyrophil cells occurring singly
or in groups throughout the bronchi and bronchioles. They
increase in number as the calibre of the airways decreases
until the terminal bronchiole is reached when the numbers fall
considerably. However occasional argyrophil cells can be found
as far out as the pulmonary alveoli (Lauweryns and Goddeeris,
1975).

Hage (1972) studied both silver reactions and APUD characteristics in the bronchi of human foetuses. No argentaffin cells were identified but argyrophil cells were found occurring singly and in clusters and in similar numbers to cells which showed APUD characteristics, including formaldehyde fume-induced fluorescence which suggests the presence of biogenic amines. Amine precursor uptake was demonstrated by prior incubation with dihydroxyphenylalanine and positive reactions were obtained with lead-haematoxylin and HCl-toluidine blue suggesting the presence of a cytoplasmic protein. These cells therefore belong to Pearce's (1968) APUD system.

By means of immunocytochemistry, Polak and colleagues (Wharton et al, 1978; Cole et al, 1980) have identified neurone-specific enolase and bombesin in these cells, the former a marker of the diffuse neuroendocrine system and the latter a fourteen amino acid peptide originally extracted from the skin of the frog Bombina. Pharmacologically, bombesin has vaso- and bronchoconstrictor properties and is a powerful releaser of other active peptides.

Lauweryns has made detailed studies of the argyrophil cells which occur in clusters. These consist of small groups of non-ciliated cells extending from the basement membrane to the lumen. The dense core granules contain serotonin and as well as a rich afferent and efferent innervation, a fenestrated capillary is found at their basal pole. Lauweryns has observed such clusters in many species including human children and adults.

In experimental animals, hypoxia induces secretion of the dense core granules at the basal (vascular) pole of the cells, a similar process to that observed in the carotid body (Lauweryns and Cokelaere, 1973; Moosavi et al, 1973). An increase in argyrophil cells has also been observed in high altitude rabbits as compared with sea-level controls (Taylor 1977). In man, Tateishi (1973) found that the argyrophil cells were most numerous when there was goblet cell hyperplasia and he suggested that the argyrophil cells may be concerned in mucus production. Alternatively hypoxia due to obstructive airway disease may underline this association of argyrophil and goblet cells. The presence of peptides as well as serotonin suggests that these cells may have various local activities, possibly modulating vasomotor tone, airway calibre, or ventilation/perfusion balance. These functions as yet remain speculative but the bronchopulmonary APUD cells may be regarded as a hypoxia-sensitive chemoreceptor system secreting its products locally and in the case of the clusters acting under neuronal influence (Lauweryns and Cokelaere, 1973). There is no evidence that the secretory products affect distant organs and the pulmonary APUD cells may subserve a paracrine rather than endocrine function, as originally proposed by Feyrter.

Endocrine tumours of the lung

Lung tumours are amongst the commonest endocrinologically active neoplasms but the endocrine activity they display may be regarded as inappropriate. The histogenesis of lung tumours associated with the various ectopic endocrine syndromes is of interest. Bronchial carcinoid tumours contain abundant dense-core granules similar to those described in APUD cells, and such granules are also consistently present in oat cell carcinoma although here they are generally quite scanty and often difficult to find. It is proposed therefore that both these tumours arise from the bronchial APUD cells, one being only locally invasive and the other highly malignant (Bensch et al, 1968). It is notable that when a pulmonary tumour is responsible for the inappropriate ACTH, ADH, calcitonin or carcinoid syndrome or for acromegaly, it is practically always a carcinoid or an oat cell carcinoma. When other types of lung tumour are associated with these syndromes, careful histological scrutiny or full necropsy examination will often lead to a revision of the original diagnosis.

Although parathyroid and placental hormone-like substances are well recognised in association with squamous, glandular and large cell undifferentiated carcinomas of the lung, these growths show no structural features suggestive of APUD cell derivation or of endocrine secretory activity. They are believed to arise from non-APUD pulmonary epithelial cells and their endocrine activity possibly reflects derepression of the genetic information common to all the body's cells.

Summary

1.   During its development the lung is a target organ for several hormones.

2.   Many substances are taken up as they pass through the pulmonary circulation. Some of these are merely bound to the lung whereas others are actively metabolised and inactivated. These substances include some hormones, but not others. The lung also activates some prohormones as they pass through the pulmonary circulation.

3.   Several bioactive substances are formed and released in
       the lung and some of these reach the systemic blood but
       it is uncertain whether this represents an accidental
       spillover or true endocrine activity.

4.   There exist in the lung certain APUD cells which
       secrete biogenic amines and peptides but whether these
       have a local paracrine or distant endocrine function is
       unknown.

5.   Lung tumours are the commonest endocrinologically
       active neoplasms, but their endocrine activity is
       "inappropriate".

REFERENCES

Ballard, P. L. and Ballard, R. A. (1979).
    Corticosteroids and respiratory distress syndrome.
    Pediatrics, 63, 163.

Bensch, K. G., Gordon, G. B. and Miller, J. R. (1965).
    Studies on the bronchial counterpart of the Kultschitsky
    (Argentaffin) cell and innervation of bronchial glands.
    Journal Ultrastructural Research, 12, 668.

Bensch, K. G., Corrin, B., Pariente, R. and Spencer, H. (1968).
    Oat-cell carcinoma of the lung.
    Cancer, 22, 1163.

Cole, G. A., Polak, J. M., Wharton, J., Marangos, P. and
    Pearse, A. G. E. (1980).
    Neuron specific enolase as a useful histochemical marker
    for the neuroendocrine system of the lung.
    Journal of Pathology, 132, 351.

Ferreira, S. H., Greene, L. J., Salgado, M.C.O. and
    Krieger, E.M. (1980).
    The fate of circulating biologically active peptides in
    the lungs.
    Metabolic activities of the lung (Ciba Foundation
    Symposium 78), 129.

Feyrter, F. (1953).
    Ueber die Peripheren Endokrinen (Parakrinen) Druesen des
    Menschen.
    W. Maudrich, Wien-Dusseldorf.

Gaddum, J. H., Hebb, C. O., Silver, A. and Swann, A. A. B.
    (1953).
    5-Hydroxytryptamine.   Pharmacological action and
    destruction in perfused lungs.
    Quarterly Journal of Experimental Physiology aand Cognate
    Medical Sciences, 38, 255.

Gryglewski, R. J., Korbut, R. and Ocetkiewicz, A. (1978).
    Generation of prostacyclin by lungs in vivo and its
    release into the arterial circulation.
    Nature, 273, 765.

Hage, E. (1972).
    Endocrine cells in bronchial mucosa of human foetuses.
    Acta pathologica Scandanavica, 80, 225.

Junod, A. F. (1972).
    Accumulation of ^{14}C-imipramine in isolated perfused rat
    lungs.
    J. Pharmacol. Exp. Ther., 183, 182.

Lauwryns, J. M. and Cokelaere, M. (1973).
    Hypoxia-sensitive neuroepithelial bodies: intrapulmonary
    secretory neuroreceptors modulated by the CNS.
    Z. Zellforsch, 145, 521.

Liggins, G. C. (1969).
    Premature delivery of foetal lambs infused with
    glucocorticoids.
    Journal of Endocrinology, 45, 515.

Moncada, S., Korbut, R., Bunting, S and Vane, J. R. (1978).
    Prostacyclin is a circulating hormone.
    Nature, 273, 767.

Moosavi, H., Smith, P. and Heath, D. (1973).
    The Feyrter cell in hypoxia.
    Thorax, 28, 729.

Nicholas, T. E., Strum, J. M., Angelo, L. S. and Junod, A. F.
    (1974).
    Site and mechanism of uptake of ^{3}HL-norepinephrine by
    isolated perfused rat lungs.
    Circulation Research, 35, 670.

Pearse, A.G.E. (1969).
    The cytochemistry and ultrastructure of polypeptide
    hormone producing cells of the APUD series and the
    embryologic, physiologic and pathologic implications of
    the concept.
    Journal of Histochemistry and Cytochemistry, 17, 303.

Ryan, U. S., Ryan, J. W., Whitaker, C. and Chiu, A. (1976).
    Localisation of angiotensin converting enzyme (kininase
    II). II. Immunocytochemistry and immunofluorescence.
    Tissue and Cell, 8, 125.

Starling, E. H. and Verney, E. B. (1925).
    The secretion of urine as studied in the isolated kidney.
    Proceedings of the Royal Society of London. B Biological
    Sciences, 97, 321.

Tateishi, R. (1973).
    Distribution of argyrophil cells in adult human lung.
    Archives of Pathology, 96, 198.

Taylor, W. (1977).
    Pulmonary argyrophil cells at high altitude.
    Journal of Pathology, 122, 137.

Wharton, J., Polak, J. M., Bloom, S. R., Ghatei, M. A.,
    Solcia, E., Brown, M. R. & Pearse, A.G.E. (1978).
    Bombesin-like immunoreactivity in the lung.
    Nature, 273, 769.

DISCUSSION

SPEAKER: CORRIN   **CHAIRMAN: CUMMING**

CHAIRMAN: The paper is open for discussion.

DENISON: Bryan, in that drawing of the microscopical
                anatomy of the lung, the coloured one, you showed
                that bronchial APUD cells are lying alongside the
                pulmonary arteries rather than veins and I
                wondered what the microscopic anatomy of well
                known endocrine organs is. Do you often find the
                secretory cells close to the arteries rather than
                being at the other side of the bed or not?

CORRIN: You usually find them very well supplied with
                capillaries and these capillaries are frequently
                lined by a fenestrated endothelium. The pulmonary
                circulation is not of the fenestrated type;
                interestingly, the bronchial capillaries do have a
                fenestrated endothelium. There are other
                differences from a typical endocrine organ and I
                think that the microscopic anatomy is more in
                keeping with my favourite hypothesis that the
                cells are paracrine rather than endocrine and that
                the pulmonary artery is not so much taking up the
                secretion as representing the target organ.

LEE: If I understand it rightly, you made the nice
                working hypothesis that this system of cells and
                the granules provided were strategically placed
                for vaso constriction, shall we say in hypoxia,
                but as I understand it this system goes on to
                produce basic amines, and basic amines are vaso
                dilators of the lung.

CORRIN: I think I showed that the cells contain the
                peptide bombesin which is probably vaso-
                constrictor in its action on pulmonary blood
                vessels. They also contain 5 hydroxytryptamine
                but I prefer to put the accent on the bombesin.

SKIDMORE: Early in your talk you mentioned the binding of
                imipramine to the lung and said that it was bound
                to the surface of the pulmonary epithelial cells;
                is that right or is it to endothelial cells?

CORRIN: It is the endothelial cell surface not the
             epithelial cell.

SKIDMORE: The lung is well known to accumulate basic
             molecules and imipramine and propranolol are good
             examples of bases that are accumulated. Do you
             think there is a general mechanism for this
             process?

CORRIN: It has been proposed that the binding capacity is
             based on the basic amine nature of many of the
             substances that are bound but I can't go further
             than that.

THERAPEUTIC AEROSOLS

Richard A. Lewis

Department of General & Thoracic Medicine

Southampton General Hospital, Southampton

Introduction

Aerosols were one of the earliest methods by which therapeutic substances were delivered to the lung. Inhalation of fumes from various sources has long been a popular treatment for asthma and the Egyptians were known to attach cones of cotton to the chest, ignite them and allow the patients to inhale the acrid smoke (Adams, 1844). More specific therapy came with the use of various herbs having atropine-like effects. One of these, **datura**, first introduced into Britain from India in 1802, was administered by inhaling smoke from the burning root of the herb, **datura ferox**. Modern therapeutic aerosol therapy followed the discovery of "adrenal substance" (Solis-Cohen, 1900) and the isolation of adrenaline (Takamine, 1902).

Administration of drugs directly to the lung by aerosol has a number of advantages. Firstly, high local therapeutic drug levels may be achieved within the airways with lower systemic concentrations and a lower incidence of consequent side-effects than other routes. Nevertheless, the respiratory tract may be a useful site for rapid absorption of soluble drugs into the blood stream. Therapeutic aerosols have been used to deliver a variety of drugs to the lung (Table 1).

Aerosol generation

Aerosols are produced by two fundamental principles; condensation of vapour molecules and comminution of macroscopic matter, (Swift, 1980). Condensation is unsuitable for

TABLE 1 — Therapeutic aerosols (Brain & Valberg, 1979)

Bronchodilator drugs

    Sympathomimetic drugs

    Xanthines

    Anticholinergic drugs

Disodium cromoglycate

Corticosteroids

Agents for oropharyngeal, laryngeal or tracheobronchial

    anaesthesia

Water or saline for humidification

Agents to aid clearance of secretions

    Detergents

    Mucolytic agents

    Proteolytic agents

Antimicrobial agents

    Antibacterial

    Antifungal

    Antiviral

Regional cancer chemotherapy

Treatment of alveolar instability with dipalmitoyl lecithin

Immunisation

The use of aerosols as a diagnostic tool is outside the scope
of this paper (see Brain and Valberg, 1979).

generation of therapeutic aerosols since the low vapour pressure of most substances used in aerosol therapy means that they are thermally degraded before sufficient vapour is available. Comminution of liquids into small particles requires energy to overcome surface tension forces. In the jet nebuliser, which is the most common generator for therapeutic aerosols, this energy is produced by compressed air. A high velocity stream of air passing over a liquid feed tube sucks liquid up the tube by the Venturi principle. The liquid is then comminuted by being drawn into sheets which are subsequently broken up. Ultrasonic nebulisers also work by comminuting liquids. In these devices a piezoelectric transducer creates high frequency pressure waves in the liquid. This causes the free surface to oscillate with the production of small surface waves (Faraday Crispations). With sufficient energy the wave crests are ruptured and elongated filaments which subsequently break into droplets escape from the liquid surface, (Boucher and Kreuter, 1968).

A third mechanism of production of aerosols is the metered dose aerosol (MDI) in which the dry drug particles are mixed with fluorocarbon propellants in a cannister at a pressure of around 3 atmospheres. These cannisters have a valve mechanism which releases a metered amount of aerosol per actuation. The highly volatile fluorocarbon propellants rapidly evaporate after release leaving smaller drug particles to enter the lung.

There are now an immense range of aerosol generators currently commercially available. Unfortunately little information is usually made available by the manufacturers about factors which affect the amount of drug delivery and the site of deposition. Such factors which would be useful for manufacturers to specify include:

1. The volume output per minute from the nebuliser.

2. The residual volume left in the nebuliser once it has been run to "dryness".

3. The optimal gas flow rate and pressure required for production of an aerosol of suitable particle size for adequate penetration into the airways.

4. The mass median diameter and distribution of particle diameters under the recommended conditions of nebulisation.

The count median diameter which is sometimes quoted by manufacturers in preference to the mass median diameter (m.m.d.) is of much less help since it bears little relationship to the amount of drug likely to be available to the lungs.

The mass median diameter from a jet nebuliser is dependent on the flow rate of driving gas. Suboptimal gas flow rates result in generation of much larger particles (Fig. 1), and an increase in variability of mass median diameter. When driving an Inspiron Mini-Neb at 10L/min we found a m.m.d. of 4.16 microns with a coefficient of variation of 0%, but a reduction in drive rate to 8L/min increased the m.m.d. to 7.97 microns with a coefficient of variation of 9%. These measurements were made using a Malvern laser particle size analyser.

Inhalation rate also affects particle size. At rapid inhalation rates larger particles, which would normally fall back into the nebuliser by gravitational forces, may be drawn up into the inspired aerosol, increasing the m.m.d. (Fig. 2).

The residual volume is the volume left in the nebuliser after a solution has been run to "dryness". Residual solution may be deposited on the walls of the nebuliser and tubing or in a pool below the feed tube. Because of the evaporative loss of water during the process of nebulisation this residual volume of solution may contain a higher concentration of solute or suspension than that originally placed in the nebuliser. The residual volume of the Inspiron Mini-neb nebuliser (C. R. Bard International Ltd.), which is the most commonly used nebuliser in hospital in the U.K. (Stainforth et al, 1982), is between 0.6 and 1.0 ml depending on how vigorously the nebuliser is shaken to displace droplets from the walls. The fact that some solution is left in the nebuliser makes the volume of diluent important. For example, (Fig. 3), if 0.5ml salbutamol respirator solution at a concentration of 5 mg/ml is diluted with 0.5 ml water or saline and run to "dryness", (i.e. to a residual volume of 0.6 ml) then 0.2 ml drug (40%) will be delivered from the nebuliser. When the volume of the diluent is increased to 5 ml then 0.45 ml (90%) of the drug is delivered from the nebuliser.

Once an aerosol has been generated and has left the nebuliser it may be affected by other factors before inhalation. The size and shape of tubing and presence of valves will influence the particle size, with larger particles depositing by inertia at bends or valves, or by sedimentation within the tubing. The use of a face mask or mouthpiece does not however appear to influence the airway response to nebulised bronchodilator aerosols (Steventon and Wilson, 1981).

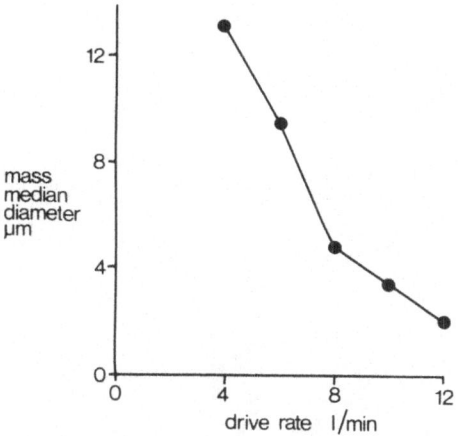

FIGURE 1   The effect of flow rate of driving gas on the
                  mass median diameter of the particles showing
                  that driving a nebuliser at too low a flow
                  rate may produce a large number of particles
                  which may be too large to enter the lung.
                  Particle size is measured by Fraunhofer
                  diffraction of laser light.

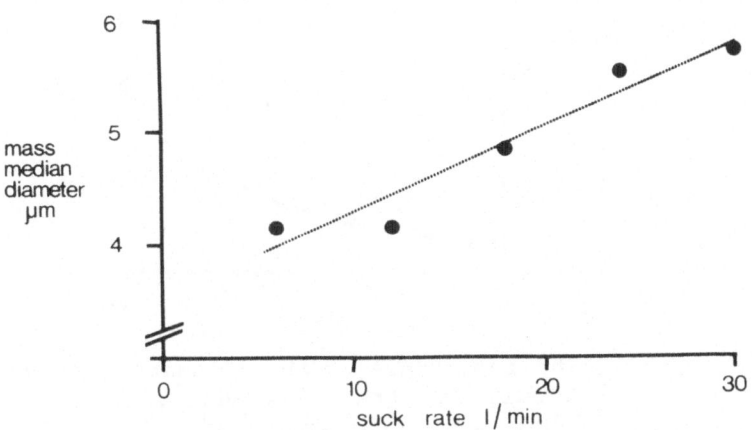

FIGURE 2   The effect of rate of suction via the mouthpiece
                  of a jet nebuliser on the mass median diameter
                  of particles from the nebulizer.

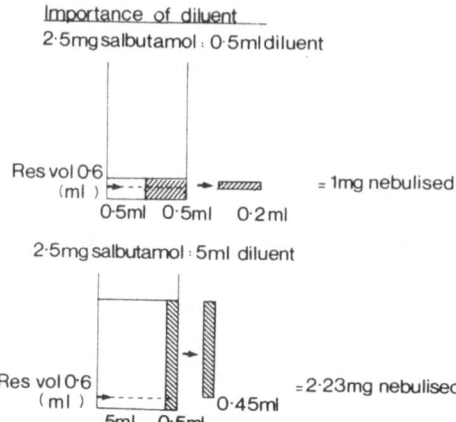

FIGURE 3   The effect of the relative humidity of the inspirate
                  from a jet nebuliser upon the mass median diameter
                  of the particles showing a rapid reduction in par-
                  ticle size as the inspirate becomes desaturated.

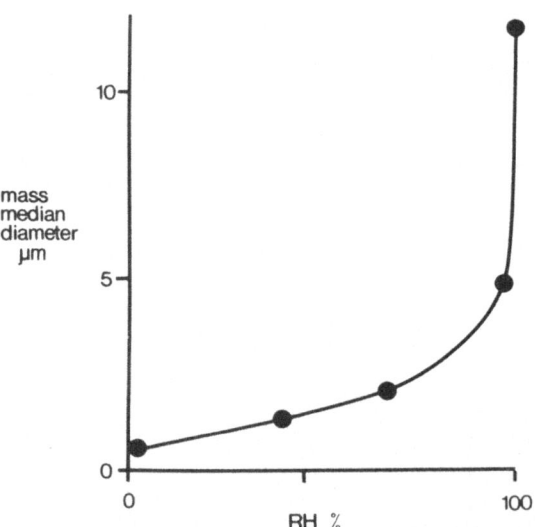

FIGURE 4   The importance of the diluent is shown for a
                  nebuliser with a residual volume of 0.6 ml
                  when run to "dryness". If 2.5 mg of
                  salbutamol respirator solution (5 mg/ml) is
                  diluted with 0.5 ml diluent and run to
                  "dryness" the 1 mg of salbutamol is delivered
                  from the nebuliser. If the same amount of
                  drug is diluted with 5 ml then 2.23 mg will
                  leave the nebuliser.

Particle size is also influenced by the humidity of the air drawn into the nebuliser to provide the auxillary air necessary for inspiration (Lewis et al 1981). When the humidity of the inspirate is unsaturated evaporation may readily take place leading to a rapid reduction in particle size (Fig. 4). We have found that in consequence there is an increase in oropharyngeal drug deposition resulting in decreased amounts of drug being delivered to the airways, and also decreased airway response to bronchoconstrictor drugs (Lewis and Tattersfield, 1980). However, the distribution of aerosol within the lungs is largely unaffected by the humidity of the inspirate. This is probably because the saline used in these studies is hygoroscopic and particles rapidly "regrow" under the saturated conditions within the airways (Scherer et al 1979).

Aerosol deposition

Deposition of an aerosol within the body results from three main mechanisms; inertial impaction, sedimentation and Brownian movement. The factors which are known to influence the deposition and response to a metered dose aerosol have recently been reviewed (Newman et al, 1981a). They found that airway response increased after slow inhalation rather than fast inhalation and a breath-holding period of ten seconds after inhalation improved response, presumably by giving time for particles to settle in the airways by sedimentation. The lung volume at which the aerosol cannister was actuated did not affect response when a 10 or 20 second breath-holding period was used. But with a shorter breath-holding period release at 20% vital capacity (VC) produced greater broncho-constriction than that at 50% or 80% VC. It is possible that with longer breath holding periods lung volume failed to have an effect because in this single-dose response study a maximal response has been achieved even with release at 80% VC. Inco-ordinated release of the cannister, either three seconds before, or immediately after inhalation significantly impaired airway response. Although Connolly (1975) found that an "open" mouth technique of inhalation improved airway response to inhaled isoprenaline this finding was not confirmed by Newman et al using more carefully controlled conditions. A summary of the most effective way to use a metered dose inhaler is shown in Table 2.

There have been fewer studies of optimal methods of inhalation from a nebuliser. Penetration of particles into the lung from an air driven nebuliser has been shown to be directly related to volume inspired per breath and inversely related to

TABLE 2 — (Newman et al, 1981a)

1.   Shake the cannister thoroughly

2.   Place the mouthpiece of the actuator between the lips

3.   Breathe out steadily

4.   Fire the inhaler while taking a slow deep inhalation

5.   Hold the breath at full inspiration for 10 sec.

TABLE 3

DRUG	RECOMMENDED DOSE (mcg)	
	MDI	NEBULISER
Ventolin (salbutamol)	200	10,000
Bricanyl (terbutaline)	500	10,000
Atrovent (ipratropium)	40	500

flow rate during inspiration (Pavia et al, 1977). Ryan et al (1981) has shown that inhalation to vital capacity over two seconds compared to eight seconds increased total lung dose and this consequently resulted in a 3.3 fold increase in PC_{20} of inhaled histamine.

A comparison of efficiency of metered dose inhaler and air driven nebuliser at lung delivery of drugs

There is a large difference in the maximum dose of drugs recommended for delivery by metered dose aerosol (MDI) compared to nebulisation (Data Sheet Compendium 1981) (Table 3). These figures may suggest that a nebuliser is less efficient at lung delivery of drug than an MDI. This impression has erroneously arisen from two sources of information on comparative efficiency of lung delivery; firstly deposition studies using radiolabelled aerosols and secondly from comparative clinical response studies. Indirect studies of deposition of (^{3}H) salbutamol were carried out using both a metered dose aerosol (Walker et al, 1972) and a variety of air driven nebulisers with and without positive pressure ventilation (Shenfield et al, 1974), measuring tritrium activity in plasma, urine and faeces. Salbutamol, if swallowed, is well metabolised in its first pass through the liver, so metabolised (^{3}H) salbutamol represents the swallowed fraction. When salbutamol is instilled directly into the bronchi free salbutamol appears in the plasma, thus free salbutamol represents the portion of drug absorbed via the lungs. With all varieties of nebuliser only 10 - 20% of the dose placed in the nebuliser was recovered from the patients, compared to 100% of the dose from an MDI. However, only 10-20% of the recovered dose from the MDI appeared to have been inhaled into the lung compared to a much larger amount from a nebuliser, since 80% of the early rise of plasma radioactivity with the nebuliser was free salbutamol. It thus appears from these studies that the initial lung fraction of delivered dose was 10 - 20% from an MDI and 8 - 16% from a nebuliser.

Subsequent direct deposition studies with both the MDI and nebuliser have confirmed that proportion of administered drug deposited in the lung is of the same order of magnitude with the two techniques. Newman et al (1981b) studied the fractional deposition of ^{99m}technetium labelled Teflon particles administered from an MDI in 8 subjects with obstructive airways disease. The results of this study are shown in Fig. 5 and compared with a study of the fractional deposition from an air driven Inspiron Mini-Neb nebuliser in 6 normal and 2 asthmatic subjects (Lewis et al, 1981a). It is clear that the percentage of the dose administered by both the

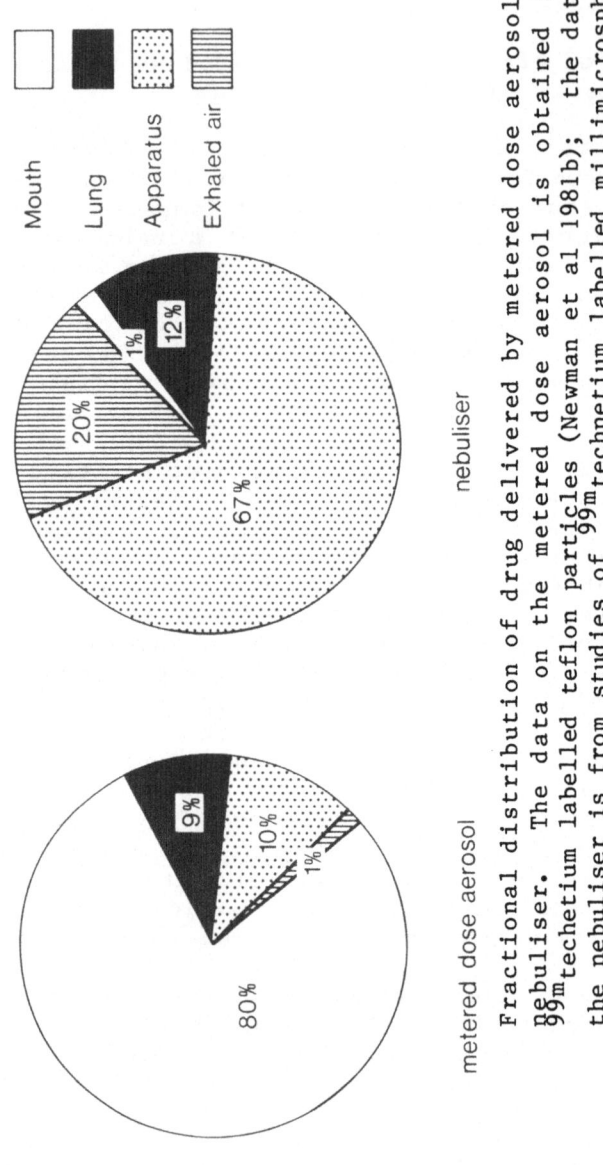

Mouth
Lung
Apparatus
Exhaled air

metered dose aerosol   nebuliser

FIGURE 5   Fractional distribution of drug delivered by metered dose aerosol and
nebuliser.   The data on the metered dose aerosol is obtained from
^{99m}techetium labelled teflon particles (Newman et al 1981b);   the data on
the nebuliser is from studies of ^{99m}technetium labelled millimicrospheres
of human serum albumin suspended in saline from an Inspiron Mini-Neb
nebuliser. In each case the figures represent the distribution leaving each
container. The amount delivered to the lungs is similar with the two
techniques.

nebuliser and MDI which subsequently deposits in the lung is of
the same order of magnitude (12% and 9% respectively). There
are, however, other large differences in fractional deposition.

        The longer length of tubing and mouthpiece of a
nebuliser is responsible for the greater amount of activity
deposited in the apparatus. There is also a large difference
in mouth deposition between nebuliser and MDI (80% and 1%
respectively) for which there are three possible explanations:
(i) the m.m.d. of nebuliser droplets is smaller. (This is
confirmed by laser measurements of particle size in our
laboratory with an m.m.d. of 2.1 micron from the nebuliser and
6.9 from the MDI); (ii) particle velocity of 30 m/sec with the
MDI, but no greater than the velocity of inspiration with the
nebuliser; (iii) particle flow within mouth and oropharynx is
fully co-ordinated with inspiration with the nebuliser but is
not necessarily co-ordinated optimally with the MDI. The
difference in particle diameter may also explain differences in
exhaled fractions.

        A number of clinical studies have compared the airway
response to salbutamol delivered by MDI and nebuliser (Table
4). In each study a small dose of salbutamol from the MDI was
compared to a much larger dose of salbutamol from a nebuliser
(17 to 50 fold larger). Despite this larger dose the airway
response was the same or only slightly greater with the
nebulisers. This might suggest that the nebuliser is less
efficient at lung delivery of drug. However, all the studies
were only single-dose studies and in this situation the
relative efficiency of various devices cannot be compared
adequately, particularly when the dose given is near to the top
of the dose-response curve. This is clearly shown in Fig. 6
where a typical salbutamol dose-response curve is seen, with
the response to 0.20 and 5.0 mg of drug indicated by dotted
lines. Because the curve plateaus at around 0.4 mg there is
little greater response from 5 mg than 0.2 mg. The response to
5 mg would be only slightly greater than the response to 0.2
mg, even if the drug was given by the same device. A full
dose-response curve to the drug delivered by nebuliser and MDI
in the same group of subjects is necessary for a true
comparison of delivery efficiency.

        When this was done in 8 normal and 8 asthmatic subjects
delivering the same cumulative dose of terbutaline by MDI and
nebuliser (Cushley et al, 1981) we found no significant
difference in the dose-response curve. This supports the
results of both direct and indirect dose deposition studies
showing that the MDI and nebuliser are equally efficient at
delivery of drugs to the lungs.

TABLE 4 – Comparative single-dose salbutamol response

Study		IPPV	Neb dose	MDI dose	Result
Choo-Kang	1975	Yes	10 mg	200 mcg	Neb MDI
Berend	1978	Yes	5 + 10 mg	200 mcg	Neb = MDI (unless severe)
Cayton	1978	Yes	5 mg	400 mcg	Neb MDI
Duncan	1979	Yes	5 – 10 mg	200 mcg	Neb = MDI
Christensson	1981	No	5 mg	300 mcg	Neb = MDI

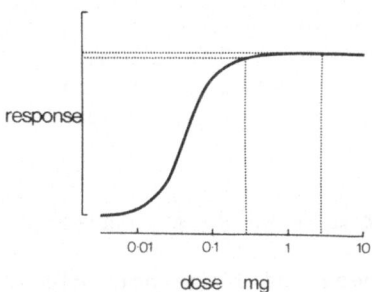

dose   mg

FIGURE 6   Typical mean dose response curve to salbutamol in
                  mild asthmatic subjects. The response to 0.2 mg of
                  salbutamol is similar to 5 mg even when the same
                  technique is used for drug delivery, i.e. metered
                  dose inhaler. Thus the finding that 5 mg delivered
                  from a nebuliser produces a similar response to
                  0.2 mg from a metered close inhaler does not mean
                  that the nebuliser is less efficient.

These findings mean that we should reconsider the recommended doses of drug for use in nebulisers since the effect of large doses of beta-agonists, particularly when given on a regular basis, has not been adequately studied. The need for vigilance, particularly when administering drugs to asthmatic patients, is indicated by past experience. Although a casual relationship between the introduction of metered dose aerosols and a rise in asthma deaths in the early 1960's has not been proven there is strong circumstantial evidence of a relationship (Inman and Adelstein, 1969). Hypotheses advanced to explain the rise in asthma deaths include cardiotoxicity of adrenergic drugs, particularly in anoxic patients, and tolerance of the airways to beta-agonist administration. Tachycardia from salbutamol aerosol has been shown to result from the inhaled rather than the swallowed fractions (Collier et al, 1980), and the recommended dose of salbutamol from a nebuliser results in up to 40 times the lung dose compared to the maximum from an MDI. This clearly indicates the need for careful studies of safety, tolerance and side effects, particularly when nebulisers are used in the home where drug dose is unsupervised.

Extension tubes for metered dose inhalers

The high speed and large particle size of droplets from an MDI results in 80% deposition by inertia in the mouth and pharynx. Extension devices were devised to slow particle velocity down by air resistance and give a greater opportunity for evaporation of the large droplets of flurocarbon propellants. Moren (1978) found that the amount of drug recovered from mouthwashing and actuator was decreased if extension tubes were placed between the lips and aerosol actuator. Such a device might also overcome the problems of poor inhaler technique. Studies of a 10 cm telescopic tube (Astra Pharmaceuticals Limited) indicated only mild benefit as assessed subjectively (Godden and Crompton, 1980), in terms of deposition (Newman et al, 1981c) or response (Bloomfield et al, 1979). One study (Gomm et al, 1980) found no improvement in response. A larger pear-shaped extension tube did show an improvement over the 10 cm spacer in terms of airway response (Lindgren et al, 1980) and lung deposition (Newman et al, 1981c). We have assessed a further modification of this pear-shaped tube which includes a one way valve at the inhalation port, the 'Nebuhaler' (Astra Pharmaceuticals Limited), in a comparative cumulative dose response study and found a marked improvement in response with significant shift to the left in terbutaline dose-response curves, both in asthmatic and normal subjects (Cushley et al, 1981). Such a device may be of particular use in inco-ordinated subjects since co-ordination is not required for its use. It could also reduce the incidence of oral candidiasis following inhalation of steroids from an MDI.

Airway response to nebulised diluent

It has long been recognised that patients with asthma and chronic chest disease complain of increasing breathlessness in an atmosphere of fog or mist. Boyd (1960), studying bronchitis mortality figures, found a strong negative correlation between mortality and temperature and absolute humidity, and less strong correlations with sulphur dioxide (SO_2) and smoke levels, the influence of SO_2 and smoke being limited to temperature below $0^{\circ}C$. Abernethy (1968), studying fog-induced bronchoconstriction, suggested that the effect was due to irritation by the water droplets in the same manner as dust particles.

Josenhans et al (1969) found that breathing humidified air caused an increase in airway resistance in asthmatic but not normal subjects. Changes in air viscosity due to changes in water vapour pressure were too small to account for any rise in airway resistance (Roy et al, 1969).

A number of workers have reported that inhalation of ultrasonically nebulised solutions of distilled water induces coughing in normal, asthmatic and bronchitic subjects, and that it induces bronchoconstriction in asthmatic and bronchitic subjects (Cheney and Butler, 1968; Pflug et al, 1970; Malik and Jenkins, 1972; Barker and Levison, 1972; Allegra and Bianco, 1980; Lilker and Jauregui, 1981; Schoeffel et al, 1981). There is less bronchoconstriction after inhalation of normal saline than after inhalation of distilled water, hypotonic or hypertonic saline (Malik and Jenkins, 1972; Allegra and Bianco, 1980; Schoeffel et al, 1981.)

Cheney and Butler (1968) also found that bronchoconstriction was not present if inhalation of saline or water was taken from a heated nebuliser at $37^{\circ}C - 40^{\circ}C$ rather than an ultrasonic nebuliser. We have since confirmed that bronchoconstriction occurs in about one third of asthmatic subjects inhaling successive solutions of normal saline from an air driven nebuliser (Lewis and Tattersfield, 1980) and that this constriction is greater when the saline is inhaled at $5^{\circ}C$ and abolished when at $37^{\circ}C$ 100% RH. This indicates that the bronchoconstriction induced by saline under these circumstances is due to a cooling effect of the inhalation on the airways, presumably by a mechanism analogous to that underlying exercise induced asthma (Deal et al, 1979).

Saline induced bronchoconstriction may also explain bronchoconstriction in asthmatic subjects attributed by Luparello et al (1968) to the effect of suggestion (Lewis et al, 1981b).

The fact that asthmatic subjects may bronchoconstrict in response to saline (Lewis and Tattersfield, 1980) or saline, sodium bicarbonate-phenol diluent (Klauster Meyer, 1979) has important implications for both use of such solutions in humidification and also for pharmacological challenge tests in these subjects. The bronchoconstriction seen after saline inhalation is progressive with successive inhalations and may not be detectable after one inhalation. Thus, in bronchial provocation tests asthmatic subjects may appear to show a dose-dependent constrictor response to increasing concentrations of drug on antigen when they are in fact responding to the cooling effect of the inhalation. A multiple dose saline control is necessary to exclude this possibility.

Acknowledgements

I wish to thank Mrs. M. Dowling for typing the manuscript, and Dr. A. E. Tattersfield for advice in all stages. I gratefully acknowledge the assistance of Dr. J. S. Fleming, Dr. W. Balachandran and Mr. M. N. Lewis in the studies described. My work was supported by a Chest Research Award from the Chest, Heart and Stroke Association.

REFERENCES

Abernethy, J. D.
Effects of inhalation of an artificial fog.
Thorax, 1968, 23, 421-426.

Adams, F.
Commentary on the seven books of Paulus Aeginatur (Trans),
London, Sydenham Society. 1844, 1, 475-478.

Allegra, L., Bianco, S.
Non-specific broncho-reactivity obtained with an ultrasonic aerosol of distilled water.
Eur. J. Resp. Dis., 1980, 61, Suppl. 106, 41-49.

Barker, R., Levison, H.
    Effects of ultrasonically nebulised distilled water on
    airway dynamics in children with cystic fibrosis and asthma.
    J. Pediatrics., 1972. 80, 396-400.

Berend, N., Webster, J., Marlin, G. E.
    Salbutamol by pressure-packed aerosol and by intermittent
    positive pressure ventilation in chronic bronchitis.
    Br. J. Dis. Chest., 1978, 72, 122-124.

Bloomfield, P., Crompton, G. K., Winsey, N. J. P.
    A tube spacer to improve inhalation of drugs from
    pressurised aerosols.
    Brit. J. Med., 1979, 2, 1479.

Boucher, R. M. G., Kreuter, J.
    The fundamentals of the ultrasonic atomisation of medicated
    solutions.
    Ann. Allergy, 1968, 26, 591-600.

Boyd, J. T.
    Climate, air pollution and mortality.
    Brit. J. prev. soc. Med., 1960, 14, 123.

Brain, J. D., Valberg, P. A.
    Deposition of aerosol in the respiratory tract.
    Am. Rev. Respir. Dis., 1979, 120, 1325-1373.

Cayton, R. M., Webber, B., Paterson, J. W., Clark, T. J. H.
    A comparison of salbutamol given by pressure-packed aerosol
    or nebulization via IPPB in acute asthma.
    Br. J. Dis. Chest, 1978, 72, 222-224.

Cheney, F. W., Butler, J.
    The effects of ultrasonically produced aerosols on airway
    resistance in man.
    Anaesthiology, 1968, 29, 1099-1106.

Christensson, P., Arborelius, M., Lilja, B.
    Salbutamol inhalation in chronic asthma bronchiale: dose
    aerosol vs. jet nebuliser.
    Chest, 1981, 79, 416-419.

Choo-Kang, Y. F. J., Grant, I. W. B.
    Comparison of two methods of administering bronchodilator
    aerosol to asthmatic patients.
    Brit. Med. J., 1975, 2, 119-120.

Collier, J. G., Dobbs, R. J., Williams, I.
    Salbutamol aerosol causes a tachycardia due to the inhaled
    rather than the swallowed fraction.
    Br. J. Clin. Pharmac., 1980, 9, 273-274.

Connolly, C. K.
    Methods of using pressurised aerosols.
    Brit. Med. J., 1975, 3, 21.

Cushley, M. J., Lewis, R. A., Cragg, D. J. B., Jackson, I. L.,
    Tattersfield, A. E.
    Administration of a beta-agonist: comparison of three
    techniques.
    Thorax, 1981, 36, 714.

Data Sheet Compendium. 1981-1982; Datapharm Publications
    Limited, London.

Deal, E. C., McFadden, E. R., Ingram, R. H., Jaeger, J. J.
    Hyperpnoea and heat influx; initial reaction sequence in
    exercise-induced asthma.
    J. Appl. Physiol., 1979, 46, 476-483.

Duncan, D., Carmichael, J., Crompton, G. K.
    Salbutamol in the treatment of asthma.
    The Practitioner. 1979, 223, 843-844.

Goddon, D. J. and Crompton, G. K.
    An objective assessment of the tube spacer in patients not
    able to use a conventional pressurised aerosol efficiently.
    Brit. J. Dis. Chest, 1981, 75, 165-168.

Gomm, S. A., Keaney, N. P., Winsey, N. J. P., Stretton, T. B.
    Effect of an extension tube on the bronchodilator efficacy
    of terbutaline delivered from a metered dose inhaler.
    Thorax. 1980, 35, 552-556.

Inman, W. H. W. and Adelstein, A. M.
    Rise and fall of asthma mortality in England and Wales in
    relation to use of pressurised aerosols.
    Lancet, 1969, ii, 279-285.

Josenhans, W. T., Melville, G. N., Ulmer, W. T.
    Effects of humidity in inspired air on airway resistance and
    functional residual capacity in patients with respiratory
    diseases.
    Respiration. 1969, 26, 435-443.

Klaustermeyer, W. B., Hale, F. C., Prescott, E. J.
    Characteristics of the asthmatic airway response to inhaled
    diluent.
    Ann. Allergy, 1979, 43, 14-18.

Lewis, R. A., Tattersfield, A. E
    Cold induced bronchoconstriction; interaction with
    prostaglandin induced bronchoconstriction.
    Clin. Sci. 1980, 59, 12P.

Lewis, R. A., Fleming,J.S., Balachandran,W., Tattersfield, A.E.
    Particle size distribution and deposition from a jet
    nebuliser: influence of humidity and temperature.
    Clin. Sci. 1981(a), 62, 5P.

Lewis, R. A., Lewis, M. N., Tattersfield, A. E.
    Psychologically induced broncho-constriction is caused by
    airway cooling.
    Thorax, 1981(b). 36, 712-713.

Lilker, E. S., Jauregus, R.
    Airway response to water inhalation: a new test for
    "bronchial reactivity".
    New Engl. J. Med., 1981, 305, 702.

Lindgren, S. B., Formgren, H., Moren, F.
    Improved aerosol therapy of asthma: effect of actuator tube
    size on drug availability.
    Eur. J. Respir. Dis., 1980, 61, 56-61.

Lippmann, M.
    Regional deposition of particles in the human respiratory
    tract.
    In: Lee, D. H. K., Falk, H. L., Murphy, S. D., ed., Handbook
    of Physiology, sec. 9, Reaction to environmental agents.
    Bethesda. Md. American Physiological Society, 1977, 213-232.

Luparello, T., Lyons, H. A., Bleecker, E. R., McFadden, E. R.
    Influences of suggestion on airway reactivity in asthmatic
    subjects.
    Psychosom. Med. 1968, 30, 819-825.

Malik, S. K., Jenkins, D. E.
    Alterations in airway dynamics following inhalation of
    ultrasonic mist. Chest. 1972. 62, 660-664.

Moren, F.
    Drug deposition of pressurised inhalation aerosols I.
    Influence of actuator tube design.
    Int. J. Pharm., 1978, 1, 205-212.

Newman, S. P., Pavia, D., Clarke, S. W.
    How should a pressurised -adrenergic bronchodilator be
    inhaled?
    Eur. J. Resp. Dis., 1981(a), 62, 3-21.

Newman, S. P., Pavia, D., Moren, F., Sheahan,N. F., Clarke,S.W.
    Deposition of pressurised aerosols in the human respiratory
    tract.
    Thorax, 1981(b), 36, 52-55.

Newman, S. P., Moren, F., Pavia, D., Little, F., Clarke, S.W.
    Deposition of pressurised suspension aerosols inhaled
    through extension devices.
    Am. Rev. Respir. Dis., 1981, 124, 317-320.

Pavia, D., Thompson, M. L., Clarke, S. W., Shannon, H. S.
    Effect of lung function and mode of inhalation on
    penetration of aerosol into the human lung.
    Thorax, 1977, 32, 194-197.

Pflug, A. E., Cheney, F. W., Bulter, J.
    The effects of an ultrasonic aerosol on pulmonary mechanics
    and arterial blood gases in patients with chronic
    bronchitis.
    Am. Rev. Respir. Dis. 1970, 101, 710-714.

Roy, P. D., Josenhans, W. T., Miller, C. H.
    Variations in air viscosity due to changes in water vapor
    pressure for isothermal conditions at temperatures below
    40°C.
    Can. J. Physiol. Pharmacol. 1970, 48, 50-53.
    Ryan, G., Dolovitch, M. B., Eng, P., Obminski, G., Cockcroft,
        D. W., Juniper, E., Hargreave, F. E., Newhouse, M. T.,

Standardisation of inhalation provocation tests: influence
of nebuliser output, particle size and method of inhalation.
J. Allergy Clin. Immunol. 1981, 67, 156-161.

Scherer, P. W., Haselton, F. R., Hanna, L. M., Stone, D. R.
Growth of hygroscopic aerosols in a model of bronchial
airways.
J. Appl. Physiol. 1979, 47, 544-550.

Schoeffel, R. E., Anderson, S. D., Altounyan, R. E. C.
Bronchial hyperreactivity in response to inhalation of
ultrasonically nebulised solutions of distilled water and
saline.
Brit. Med. J., 1981, 283, 1285-1287.

Shenfield, G. M., Evans, M. E., Paterson, J. E.
The effect of different nebulisers with and without
intermittent positive pressure breathing on the absorption
and metabolism of salbutamol.
Br. J. Clin. Pharmac. 1974, 1, 295-300.

Solis-Cohen, S.
The use of adrenal substance in the treatment of asthma.
J.A.M.A. 1900, 34, 1164-1166.

Stainforth, J. N., Lewis, R. A., Tattersfield, A. E.
The use of nebulisers to administer beta-agonists in
hospital: a questionnaire study.
Thorax. 1982 (in press).

Steventon, R. D., Wilson, R. S. E.
Facemask or mouthpiece for delivery of nebulised
bronchodilator aerosols?
Br. J. Dis. Chest. 1981, 75, 88-90.

Swift, D. L.
Aerosols and humidity therapy. Generation and respiratory
deposition of therapeutic aerosols.
Am. Rev. Resp. Dis., 1980, 122, 71-77.

Takamine, D. J.
Adrenalin - the active principle of the suprarenal gland.
Scott. Med. and Surg. J. 1902, 10, 131-138.

Walker, S. R., Evans, M. E., Richards, A. J., Paterson, J. W.
The clinical pharmacology of oral and inhaled salbutamol.
Clin. Pharmac. Ther. 1972, 13, 861-867.

DISCUSSION

SPEAKER: LEWIS   **CHAIRMAN: CUMMING**

CHAIRMAN: I am sure that this presentation has stimulated
             many questions The paper is now open for
             discussion.

CORRIN: Some years ago Newhouse from Ontario, Canada
             suggested that it would be advantageous to give
             different drugs into the lung in different size
             aerosols in order to optimize the deposition of
             the drugs in the airways where they would work
             best. Do you know if he has followed up that
             suggestion and got evidence that it is so, or
             whether it really doesn't matter providing the
             aerosols reach the lung which generation of airway
             the particles land in.

LEWIS: Yes, he did follow up the suggestion and the group
             have published a paper (Ryan 1981) in which they
             assessed the effect of particle size on the airway
             response to inhaled methacholine. They found that
             nebulizer output, but not particle size affected
             the response. The problem with this study is that
             they gave polydisperse aerosols in which there was
             some overlap in particle size, so the
             discrimination they were able to obtain in that
             study was not as good as one would have hoped to
             be able to obtain using discretely sized
             monodisperse aerosols. Work needs doing along
             these lines, but I am sure some groups are
             carrying it out at present.

GUINTINI: You didn't show data regarding the intrapulmonary
             distribution of aerosols generated by the metered
             dose inhaler and ordinary jet nebuliser. Would
             you like to say something about that.

LEWIS: Yes, one way of assessing the distribution of
             aerosols within the lung is to rescan after 24
             hours after inhalation when aerosol deposited in
             the conducting airways has been cleared and what
             is left after correction for decay of
             radioactivity represents the deposition in non-
             conducting airways. Comparing my own studies with

a jet nebuliser to Steven Newman's with the metered dose inhaler, the distribution between conducting and non-conducting airways is similar.

GUINTINI: Similar, you mean, to the aerosol of size range in the order of about 3 microns which is almost entirely deposited in the first, maybe, 6 to 8 generations.

LEWIS: I am afraid that I do not know of any way of telling which generation it is deposited into in intact man. All I know is what is deposited in the non-conducting compared to the conducting airways.

GUINTINI: You know that when you are dealing with a particle which hs about 1 micron mass median diameter the particle taken as reference is going to be deposited in the lung mostly by sedimentation which means that the majority of these particles is deposited in the peripheral airways. So when you are dealing with particles with a mass median diameter of 3 - 4 microns, as usually delivered by metered dose inhaler, the deposition by impaction is greater than the deposition by sedimentation, and this has been shown in in normal human airways. So I am always surprised when I listen to such a similarity of deposition, because, of course, you can produce from an ordinary jet nebuliser any kind of size distribution, and certainly you can obtain a deposition from a jet nebuliser that is going to match that of a metered dose nebuliser. But you can also do much better using a jet nebuliser.

LEWIS: Two points. One is that the most important parameter which affects deposition is the rate and volume of inhalation of each subject. Because all subjects inhale differently this makes intersubject variation very great. Had we used exactly the same technique of inhalation for each subject, then perhaps we would have been a little more discriminating. As far as the second point, impaction, is concerned, I think, with the metered dose inhaler a lot of those large particles, the ones say of about 6 - 8 microns, were impacting in the oro-pharynx where 80% of the activity impacted. So that particles actually getting down into the airways would have represented much smaller particles. A fairly wide distribution of particles is delivered by the metered dose inhaler. We also know that with the MDI 80% of particles deposit at the mouth compared to 2% with the nebuliser.

CHAIRMAN: Can you say anything about the charge carried by the particles from the jet nebuliser?

LEWIS: They carry a very high electric charge. In our experiments the charge has been found to be as high as 2 Coulombs/m^2.

CHRONIC EFFECTS OF DRUGS ON AIRWAY MUCUS-SECRETING CELLS

P. K. Jeffery, D. F. Rogers, M. Ayres, P. M. Evans & D. A. William

Department of Lung Pathology, Cardiothoracic Institute

Brompton Hospital, London

Atmospheric pollution and, in particular, personal pollution by tobacco smoke (T.S.) is a major cause of chronic inflammation of the tracheo-bronchial mucous membrane (Royal College of Physicians, 1977). In man there is an increase in the number of mucus-secreting cells both in the surface epithelium and in submucosal glands (resulting in gland hypertrophy). The clinical consequences are recurrent cough and the excessive production of airways' mucus which is expectorated as sputum (Reid 1954). Small airways, normally without mucus-secreting cells, are also affected and show secretory cell metaplasia, a bronchiolitis and smooth muscle hypertrophy (Hogg et al, 1968; Cosio et al, 1980). Thus whilst the contribution to secretory cell increase by surface and gland cells is greatest in the largest airways, the secretory cell metaplasia seen in the epithelium of bronchioli probably has the greater functional implications (Hogg et al, 1968; Thurlbeck et al, 1975).

Basically three cell types make up the secretory cell population: mucous and serous cells present both in the submucosal glands (Lamb & Reid 1969, 1970; Meyrick & Reid, 1970) and surface epithelium (Jeffery 1982; Jeffery and Reid, 1975; 1977) and Clara cells which are normally the only secretory cell type present in distal bronchioli (Plopper et al, 1980).

The mucous and serous cells of the surface epithelium (previously called "goblet cells") are most easily distinguished by electron microscopy (EM). The serous cell has electron-dense cytoplasm, rough endoplasmic reticulum and

87

sparse numbers of discrete, electron-dense secretory granules contrasting with the larger electron-lucent and often confluent granules of the mucous cell. The bronchiolar Clara cell may resemble the serous but unlike the latter contains abundant smooth endoplasmic reticulum (Jeffery & Reid, 1975). In healthy man the mucous and serous cells are limited to cartilagenous airways but in disease they may extend into bronchioli and then increase in number.

Submucosal glands are predominantly composed of mucous and serous cells whose EM morphologies resemble those described in the epithelium. Secretory acini may be composed only of one or a mixture of these cell types. Each acinar tubule connects with, and its mucinous contents pass into, a collecting duct region which passes secretion via a ciliated duct to the airway lumen (Meyrick et al, 1969). In healthy man the gland mass is restricted to cartilagenous airways and occupies about one third the depth of the airway wall whilst in hypersecretory diseases it may increase to two thirds of the wall (Reid, 1960). The amount of gland and its predominant site varies greatly with species (Goco, 1963) and sex differences have been found (Hayashi et al, 1979).

The increase in mucus-secreting cells seen in disease can be mimicked experimentally by passively exposing specific pathogen-free (SPF) rats to TS or SO_2 for as little as two to three weeks. There is gland hyperplasia and surface mucous cell hyperplasia: the latter is found in both extra pulmonary and intrapulmonary airways (Reid, 1963; Lamb & Reid, 1968; 1969; Jones et al, 1972; Jeffery & Reid, 1977 & 1981). The rat model of hypersecretion as well as those developed for the dog (Spicer et al, 1974), pig (Baskerville, 1976) and cat (Kleinerman et al, 1976) now give the experimental pathologist/physiologist a means by which the effects of drugs on the amount of mucus-secreting tissue may be investigated.

Whilst the acute effects of drugs* on the discharge of mucus from airway mucus-secreting cells have been reviewed by Richardson & Phipps (1978) and Phipps (1981) the chronic effects have not. Accordingly the present paper reviews:

---

* Defined as any substance used as a 'medicine' to cure, heal or relieve, or showing a pharmacological effect (i.e. specifically blocked) or having a specific biochemical/biological activity.

(1) the drugs which when given repeatedly increase the amount of mucus-secreting tissue (as determined by histology),

(2)   the effects of anti-inflammatory and mitotic-arresting (stathmokinetic) drugs in inhibiting the increase in mucus-secreting cells seen after irritation by T.S., and

(3) the possible beneficial effects of two anti-inflammatory drugs in shortening time taken to recover following cessation of T.S. exposure.

Following this there will be a short fourth section on mechanisms of action.

1) DRUGS WHICH INCREASE MUCUS-SECRETING TISSUE

Table 1 shows 5 pharmacological agents, one hormone and one enzyme which have been shown to cause an increase in secretory cell number experimentally when given either subcutaneously (sc), intraperitoneally (ip), intramuscularly (im), orally or by aerosol (aer) to animals.

Isoprenaline (isoproterenol or isopropylnorepinephrine)

Isoprenaline (here referred to as IPN) is one of the most active sympathomimetic amines and is a selective B adrenergic agonist which has both B_1 and B_2 effects. IPN is employed clinically as a bronchodilator or as a cardiac stimulant in heart block, cardiogenic shock after myocardial infarction, or septicemic shock (Goodman & Gilman, 1970). Experimentally it has been used in 'high' doses of 40 or 100 mg kg daily for 6 or 12 days: s.c. to rats and has been shown to cause an increase in the number of surface secretory cells and a hypertrophy of submucosal glands (Sturgess & Reid, 1973; Jones & Reid, 1979; Reid & Jones, 1980). The IPN effect was apparent even under germ free conditions and also caused a hypertrophy of pancreatic and salivary glands. In the heart it caused an increase in ventricular weight (the right more than the left) and some myocardial damage. The effect on secretory cell number was asymmetric in that the increase was due to a selective change in those cells containing acidic glycoprotein with little change in those containing neutral glycoprotein. In each case the secretory cells were more distinct due to increases in the amounts of intracellular secretion and in cell size. The increase in submucosal gland width, length and depth were greater and higher then dose and number of injections. The effect on secretory cell number was present even after one injection and was marked in the trachea and at each

inrapulmonary airway level studied. By EM Jeffery (1973) has
shown that the increase seen in the bronchiolus may take place
as the result of a transformation of a number of cell types
including the Clara, ciliated and brush cells (fig. 1, 2 & 3).

IPN has also been found to produce similar changes in
minimal disease pigs. Following injections of 4 mg/kg (im)
daily for 6 consecutive days, Baskerville (1976) found a
similar hypertrophy of submucosal glands and a shift in the
intracellular glycoprotein to the acidic (sialic acid) type.
There was also an increase in epithelial secretory cell number
but no change in their sialylated glycoprotein. The heart
(both ventricles) and the salivary and adrenal glands were
hypertrophied also.

Salbutamol (albuterol)

Salbutamol (referred to here as SBM) is reputed to be a
sympathominetic with high selectivity for B_2 receptors and is
used clinically in the treatment of bronchial asthma to relax
selectively bronchial smooth muscle without the undesirable
cardiac stimulation seen with IPN (Choo-Kang et al, 1970).
Given at the same high dose as IPN (100 mg/kg as sulphate), SBM
had a lesser effect and only caused a significant rise in
secretory cell number at two airway levels (Reid & Jones 1980).
Unlike IPN no increase in bronchiolar mucus cells was detected
by EM (author's unpublished data).

Pilocarpine

Pilocarpine (referred to here as PCP) is a naturally
occurring alkaloid and cholinomimetic which stimulates
autonomic effector cells acted on by postganglionic cholinergic
nerves. Used clinically as a pupillary constrictor, it is also
known to cause contraction of bronchial smooth muscle and the
discharge of intracellular mucus from bronchial submucosal
glands but not surface epithelial secretory cells (Florey et
al, 1932; Sturgess and Reid, 1972; Gallagher et al, 1975).
However, PCP (4 mg/kg, s.c. as nitrate for 6 or 12 days to
rats) increases both the number of epithelial secretory cells
and the size of submucosal glands. In contrast to IPN, 12 days
of PCP injections increases the number of both neutral and
acidic secretory cells. Whilst the overall gland size was
increased after 12 days, the secretory cells appeared to be
exhausted of their secretion. This feature was also repeated
in the pancreas and salivary glands (Sturgess & Reid, 1973).
EM of animals selected at random from these experiments failed
to confirm the results and showed instead an increase in the

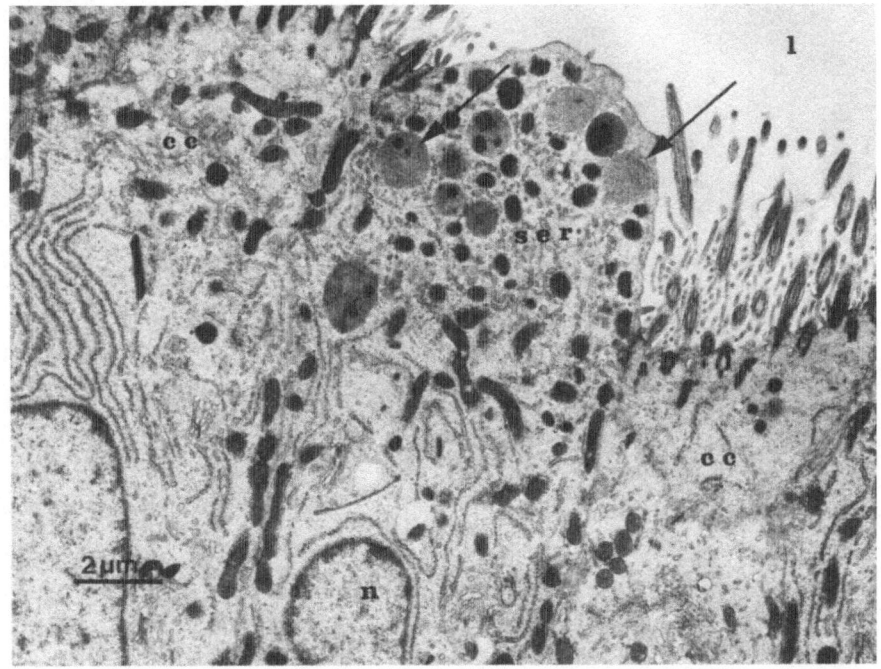

FIGURE 1: An electron micrograph of a bronchiole from a rat given isoprenaline sulphate daily for 12 days. Bordered by two ciliated cells (cc), the central cell has an abundance of smooth endoplasmic reticulum (ser), a characteristic of Clara cells, but also large electron-lucent secretory granules of the type commonly found in mucous cells (arrows). Cell nucleus (n) and airway lumen (l). Glutaraldehyde plus osmium tetroxide: uranyl acetate and lead citrate. 2 um marker, bottom left.

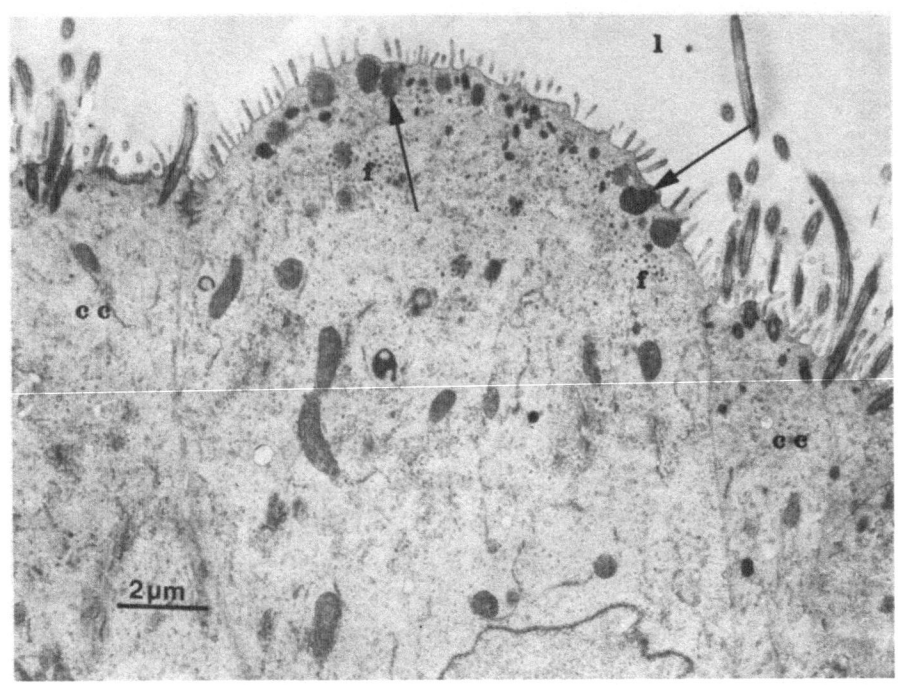

FIGURE 2:   Electron micrograph from an animal similarly
             treated. The apex of the central cell has an
             electron-lucent cytoplasm, fibrogranular
             accumulations (f) and regularly-spaced luminal
             microvilli typical of ciliated cells (cc), whilst
             at the same time the cell contains secretory-like
             granules (arrows).
             Glutaraldehyde and osmium tetroxide: uranyl
             acetate and lead citrate. 2 um marker.

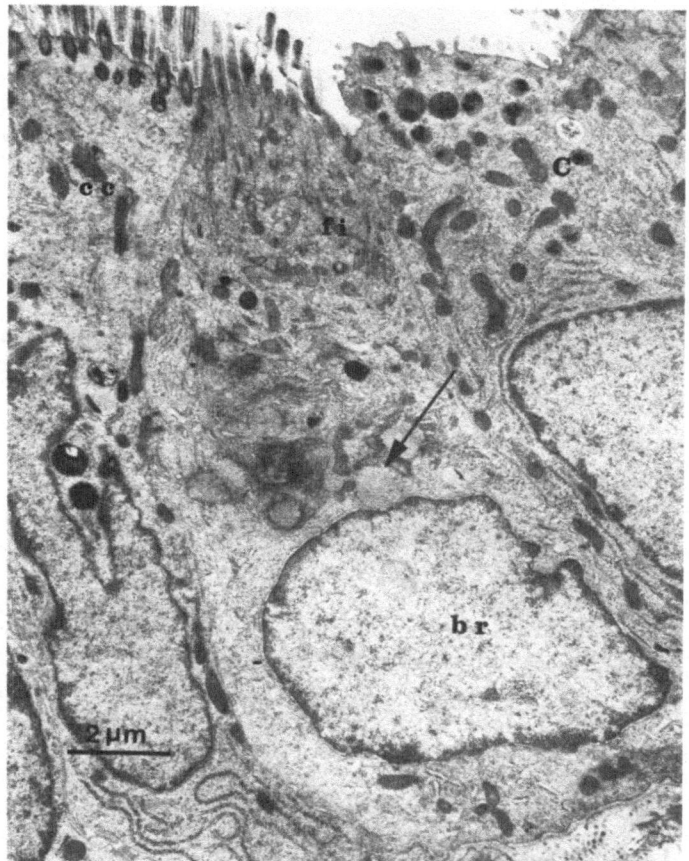

FIGURE 3:   Electron micrograph of an isoprenoline-treated
              animal illustrating a brush cell, recognisable by
              its brush border and bundles of filaments (fi).
              Secretory-like granules are additionally present
              in the supra-nuclear zone (arrow). Ciliated cell
              (cc), Clara-like cell (C) and nucleus of brush
              cell (br).
              Glutaraldehyde and osmium tetroxide: uranyl
              acetate and lead citrate. 2 um marker.

number of cells without secretory granules many of which gave
the impression of recent evacuation of secretory content
(Jeffery, 1973).

Methacholine (acetyl-B-methycholine)

The synthetic choline derivative methacholine (referred to
here as M.Ch.) has a greater duration and selectivity (i.e.
preferential muscarinic rather than nicotinic) of action than
acetylcholine and may be used as a vasodilator, cardiac
vagomimetic or as a pupillary constrictor. Kleinerman et al
(1976) used it experimentally to induce the histological
hallmarks of human chronic bronchitis in cats. Daily
administration (3 mg/kg im) over 90 days resulted in an
increase in bronchial gland mass and a hypertrophied epithelium
in which there was an increase in secretory cell number.
Bronchial explants cultured from these animals showed an
increase in total sialic acid recovered in a 24 hour mucin
collection.

Nicotine

As a naturally occurring alkaloid first isolated from the
leaves of the tobacco plant Nicotiana tobacum, nicotine is of
interest in regard its potentially harmful role in tobacco
smoke-induced disease and as a pharmacological tool used to
stimulate autonomic ganglia, neuromuscular junctions, and
sensory receptors (e.g. mechanoreceptors of lung). Its major
action is transient stimulation ('low' doses) with subsequent
persistant depression of all autonomic ganglia, including the
adrenal medulla ('high' doses). The same biphasic response is
seen in the discharge of secretion by salivary and bronchial
glands (Nedergaard & Schrold, 1977; Goodman & Gilman, 1975).

In order to investigate the role of nicotine in tobacco
smoke-induced secretory cell hyperplasia L-nicotine tartrate
was given by injection (sc) to rats at a 'low' (20 mg/kg daily
for 22 days producing a mean plasma nicotine level of 650
ng/ml) or 'high' dose (8 x 7.5 mg per day for 14 days; plasma
concentration of 2350 mg/ml and by aerosol (0.5 mg/ml
maintained for 2 hours/day for 22 days: plasma concentration 10
ng/ml) (Roger et al - in preparation). The plasma nicotine
concentrations found in rats given a regime of whole tobacco
smoke known to cause secretory cell hyperplasia has been found
to be 100 ug/ml (ibid.). The results for the counts of
epithelial secretory cell numbers shows that the 'low' dose
caused a reduction in secretory cell number but only in
intrapulmonary airways. However, the 'high' dose caused an
increase in secretory cell number at all airway levels.

Table 1. Drugs Which Affect Bronchial Mucus-Secreting Tissue

Drug	Clinical Interest	Epithelial cells Hyperplasia	Hypertrophy	Sub-mucosal Glands Hypertrophy
Isoprenaline	Bronchodilator	⬆ (large)	–	↑ (small)
Salbutamol	Bronchodilator	↑ (small)	–	–
Pilocarpine	Pupillary constrictor	↑ (small)	–	↑ (small)
Methacholine	Vasodilator	⬆ (large)	↑ (small)	↑ (small)
Nicotine	Constituent of tobacco	⬆ (large)	–	–
Elastase	Proteolytic enzyme	⬆ (large)	↑ (small)	–
Oestrogens	Contraception	⬆ (large)	↑ (small)	–

⬆ = large increase   ↑ = small increase   – = not documented

= only at 'high' dose (see text)

Table 2. Potency of Inhibition By Anti-Inflammatory Drugs*
(based upon statistical comparison of mucous cell
numbers).

Airway	Indomethacin	Dexamethasone	Prednisolone	Hydrocortisone
TRACHEA	***	*	**	*
AXIAL (Upper and lower)	***	***	***	***
BRONCHIOLI				
Proximal				
upper	***	**	–	*
lower	***	*	*	**
Distal				
upper	**	***	**	–
lower	**	***	**	–
Total star score (out of a possible 18)	16	13	10	7

* efficacy of drug   *** = most   ** = second most   * = third most   – equals least effective

Exposure to the nicotine aerosol had no effect on secretory
cell number. The 'high' dose was the only nicotine treatment
to have a detrimental effect on the health of the rats.
Nicotine is thus unlikely to have been a major determinant of
T.S.-induced secretory cell hyperplasia in our experimental
studies.

Oestrogens

Several ovarian oestrogens are produced in a cyclical
fashion with oestradiol 17B being the most active and the one
responsible for proliferation of vaginal and uterine mucosae
and for the increase in secretion by cervical glands. The
discovery that stilbene derivatives had similar physiological
effects, when given orally, had led to their use as
constituents of some oral contraceptives.

Quantitative cytology of airway brushings has shown that
secretory cells exhibit cyclical changes in their numbers
during the menstrual cycle, reaching a maximum 33% at the end
of the cycle and reduced to 10% on day 13 (Chalon et al 1971).
Cyclical changes in hormone levels may be responsible for this
effect and there is supportive evidence from animal studies.
For example, Hayashi & Huber (1977) studied the secretory cell
numbers in the trachea of male rats and compared them with
those found in female rats at distinct phases of their cycle.
Whilst the tracheal epithelium of male rats was the thicker,
female rats always had a higher epithelial secretory cell
number. Oestrus and proestrus rats contained significantly
more secretory cells than did dioestrus rats when the mean
value approached that found in males. These differences were
mirrored when the cells containing neutral glycoprotein were
compared separately. Those cells with intracellular acidic
glycoprotein were consistently lower in the female than male
rats. Following exposure to whole TS for 30 consecutive days
(Hayashi et al, 1978) the increase in secretory cells due to
smoke was greater in female than in male rats. There was also
a differential shift in glycoprotein type with cells containing
neutral glycoprotein being increased in the females but not
males. The airways of male rats exposed to TS contained a
substantially higher proportion of cells with acidic
glycoprotein than did females. The authors postulated that the
higher proportion of cells containing acid glycoprotein and
greater epithelial thickness of males may be anatomic features
important in the pathogenesis of chronic bronchitis in men.
Whether or not these differences are due to oestrogens or other
female hormones is unclear although the former is strongly
implicated. There is supportive experimental evidence from the
studies of El-Heneidy et al (1966) and El-Ghazzawi et al

(1979). Chronic administration of ethanyl oestradiol (25 mg. orally, daily) given to guinea pigs has been found to cause initial increases in the number of airway secretory cells which are subsequently replaced by epithelium of the stratified squamous type.

Elastase (E.C.3.4.4.7)

Intratracheal instillation of proteolytic enzymes induces emphysematous changes in the lungs of experimental animals. A single instillation of elastase (2 mg/kg) to hamster trachea produces an emphysema which is pan acinar (panlobular) in type (Hayes et al, 1975) and produces an airway mucosal lesion which resembles the mucous cell metaplasia often described in this condition in man (Christensen et al, 1977; Hayes & Christensen, 1978). Bronchial changes were followed up to 24 days post exposure when the secretory cell number and hypertrophy reached a peak. As the mucous 'blanket', seen by scanning EM was more complete in the elastase-treated than in the untreated animals, the authors suggested that there was concomitant hypersecretion as well as hyperplasia. Longer term studies indicated that the lesion does not revert to normal within a year which highlights the uniqueness of this form of injury.

2. INHIBITION OF T.S.-INDUCED CHANGES

a) Anti inflammatory drugs

Following the discovery that the anti-inflammatory drug phenylmethyloxadiazole (PMO given orally, as oxalamine citrate or as 2% by weight of cigarette tobacco) would attenuate the ciliostatic effect of T.S. in in situ, segments of cat trachea (Dalhamn, 1966; 1969; 1971; Dalhamn & Rylander, 1971), in vivo animal studies have demonstrated that, as 2% in tobacco, PMO also had an inhibitory effect on the secretory cell increase seen after exposure to T.S. alone (Jones et al, 1972; 1978; Jeffery & Reid, 1981). As the mechanism of action of PMO is little understood, these experimental studies have been extended to examine the effects of other anti-inflammatory drugs whose mechanisms of action are better documented.

The non-steroidal drugs indomethacin and Flurbiprofen have been given by i.p. injection at a range of doses between 0.4 to 4 mg/kg daily and concurrently with exposure of rats to T.S. Their effective inhibition has been determined and compared with that of the steroid drugs dexamethasone, prednisolone, and hydrocortisone all given at 4 mg/kg (Rogers and Jeffery, 1981; Jeffery et al, 1982). Comparison (table 2) shows that while

all inhibit the increase in secretory cell number due to T.S.
alone, indomethacin is the most effective overall. The degree
of inhibition by the steroidal drugs correlates well with their
claimed order of potency against more classical models of
inflammation such as paw oedema (Sayers & Travis, 1970). The
effective inhibition by indomethacin is dose related and is
least in the trachea and greatest in the most distal airway
studied (Greig et al, 1980). This contrasts with the finding
that PMO has its greatest effect in the trachea and has little
effect on the TS-induced increase seen in intrapulmonary
bronchi (Jones & Reid, 1978). Flurbiprofen was the only drug
found to have an inhibitory effect when given orally.

b) Vinblastine (vinca leukoblastine)

    Vinblastine is one of the naturally occurring vinca
alkaloids derived from the periwinkle plant (Vinca rosea Linn),
and has been used clinically in the treatment of various forms
of cancer and in patients with Hodgkin's disease. The vinka
alkaloids are cell-cycle-specific agents and block mitosis at
metaphase, probably due to dissolution of microtubules
necessary for continuation into telophase (Goodman & Gilman,
1975).   Recent experiments have shown that an increase in
mitoses is one of the mechanisms responsible for the ensuing
secretory cell increase (Ayres & Jeffery, 1982; Jeffery et al,
1982). Preliminary results of experiments designed to see if
Vinblastine might therefore inhibit secretory cell hyperplasia,
have shown that it inhibits TS-induced secretory cell
hyperplasia. However, the effect of Vinblastine did not appear
to be mediated by inhibition of mitosis as at the dose used
chronically (0.05 mg/kg) there was no accumulation of metaphase
arrests with time (Evans et al unpublished) suggesting an
alternative mechanism of action.

3. REVERSIBILITY AND RECOVERY

    A number of studies in man now show that cessation of
smoking results in subjective and objective clinical
improvement. For example:

a) Pulmonary epithelial permeability is increased in otherwise
   healthy smokers. There is significant improvement in this
   defect after only 24 hours with a maximum improvement by 7
   days following cessation of smoking. The maximum recovery
   value is, however, still below that for non-smokers (Minty
   et al, 1981).

b) Tracheobronchial clearance is reputed to be reduced in

chronic smokers (Camner & Philipson, 1972). Following cessation of smoking, clearance was not improved by one week but returned to near control values by three months (Camner et al, 1973).

c) Experimental studies show that recovery of secretory cell hyperplasia, induced by exposure of rats to T.S. (2 weeks), takes between 9 and 42 days depending on the airway level studied. (Rogers & Jeffery – in preparation).

Recovery of IPN-induced mucous cell hyperplasia in the pig takes longer and is complete by 4 weeks for submucosal glands and 8 weeks for epithelial secretory cells (Baskerville, 1976).

Following cessation if IPN or SBM given to rats, secretory cell numbers in the trachea return to normal within about a week whilst in other regions (e.g. main bronchus) it may persist for up to 12 weeks (Reid & Jones, 1980).

The question as to whether drugs may reduce the time taken for secretory cell numbers to return to normal has only recently received attention (Rogers & Jeffery – in preparation). To date results are available for two non-steroidal anti-inflammatory drugs: namely indomethacin and Flurbiprofen given (i.p.) as 4 mg/kg daily to rats following cessation of tobacco smoke. Both drugs significantly increase the rate of return to control values which in the distal airways are reached within as little as 4 days. This is the first such demonstration experimentally and has obvious clinical implications.

4. MODE OF ACTION

There are several ways in which the drugs, cited here, might be acting either to induce secretory cell hyperplasia or, alternatively, to inhibit the hyperplasia due to T.S.

It should be remembered that detection of mucus-secreting cells by light microscopy of paraffin-embedded sections is dependent on the identification of intracellular mucin by specific stains. Once a secretory cell has completely discharged its content it will not be visible and there will be an "apparent" reduction in secretory cell number. Thus the presence and amount of detectable intracellular ('storage') mucin relies on the balance of mucin-precursor uptake and discharge and the level set for this balance (fig. 4).

Secretory cell hyperplasia could therefore be due to:

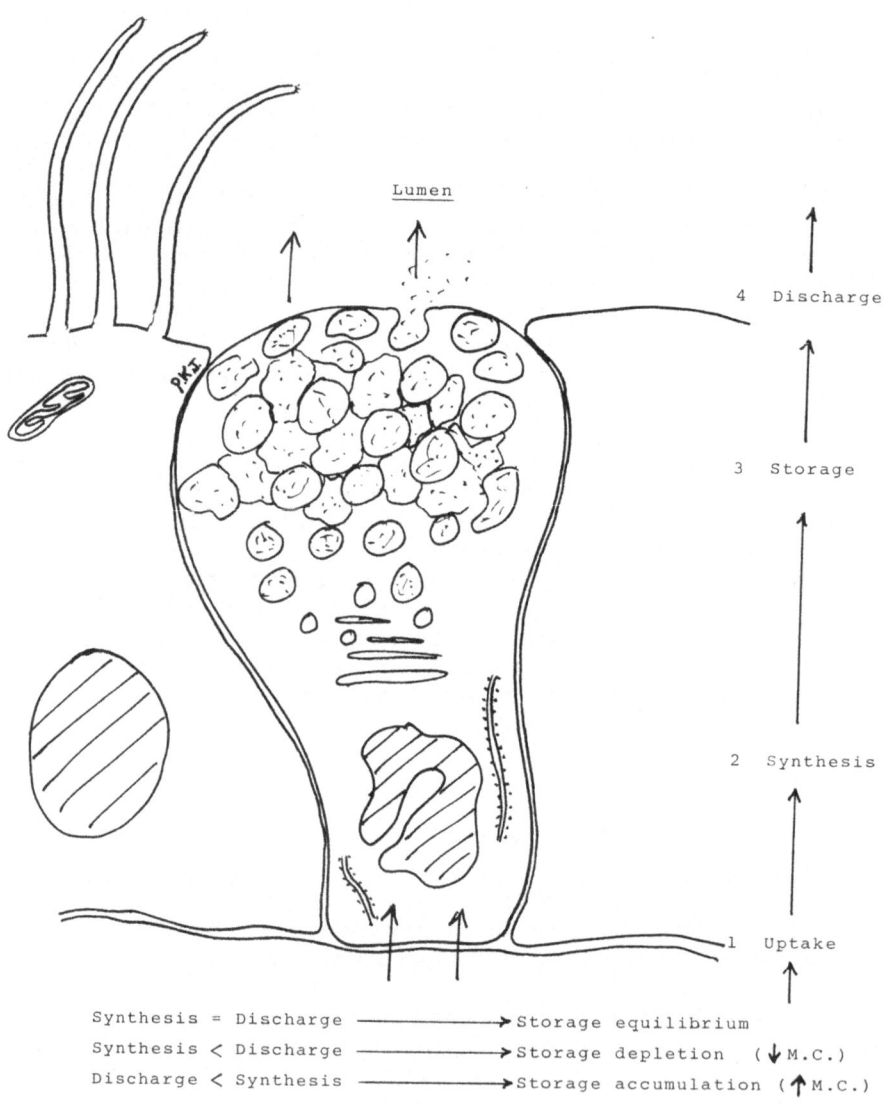

Synthesis = Discharge ──────────► Storage equilibrium
Synthesis < Discharge ──────────► Storage depletion (↓M.C.)
Discharge < Synthesis ──────────► Storage accumulation (↑M.C.)

FIGURE 4.

a)   an increase in the amount of intracellular mucus by either decreased discharge (Adler et al, 1981) or increased synthesis.

b)   stimulation of cell proliferation (Ayres & Jeffery, 1982; Jeffery et al, 1982).

c)   transformation of serous, Clara or other cell types to mucous cells (Jeffery, 1973; Jeffery & Reid, 1981; Jeffery et al, 1982). All the above may be mediated by cyclo-oxygenase or lypoxygenase metabolites of arachidonic acid. For example, both nicotine and TS act to increase the tissue levels of prostaglandins in the lung (Berry et al, 1979; Bahkle et al, 1979) and these could have secondary effects on mucin synthesis and release (Mahoney & Waterbury, 1981; Marom et al, 1981) and on cell proliferation (Sykes & Maddox, 1972; Humes & Strausser, 1974; Abell & Monohan, 1978).

Similarly the inhibition of TS-induced secretory cell hyperplasia by anti-inflammatory drugs could be due to inhibition of:

a)   mucin synthesis (Musil et al, 1968; Lukie & Forstner, 1972; Coles et al, 1979.)

b)   cell proliferation (Bayer & Beaver, 1979).

c)   prostaglandin synthesis and release and thereby their secondary effects (Vane, 1971; Fereira, 1972; Lewis & Piper, 1975; Moncada, 1979; Flower, 1974; 1978; 1979).

CONCLUSIONS

A number of drugs have a long-term effect on the amount of mucus-secreting tissue in the respiratory tract. We have reviewed the effects of drugs in:

1.   promoting mucous cell hyperplasia. In particular isoprenaline, methacholine, nicotine, oestrogen-like compounds and elastase and to a lesser extent salbutamol and pilocarpine are active.

2.   inhibiting tobacco smoke-induced changes. Both non-steroidal and steroidal anti-inflammatory drugs and, in particular, indomethacin are effective.

3.   reducing the time taken to recover following cessation of
     tobacco smoke exposure Indomethacin and Flurbiprofen are
     effective.

   These effects could be produced in several ways currently
under investigation.

ACKNOWLEDGEMENTS

   We are grateful for the support given to some of these
studies by the Medical Research Council of Great Britain.,
Wellcome Trust and Cystic Fibrosis Research Trust. We also
thank C. Feyerabend, New Cross Hospital, London, England for
the plasma nicotine estimations.

REFERENCES

ABELL, C. W., Monahan, T.M. (1973),
   The role of adenosine 3', 5'-cyclic monophosphate in the
   regulation of mammalian cell division.
   J. Cell Biol., 59, 549.

ADLER, K. B., Brody, A. R., Craighead, J. E. (1981).
   Studies on the mechanism of mucin secretion by cells of the
   procine tracheal epithelium (41030).
   Proc. Soc. Exp. Biol. Med., 166, 96.

BAHKLE, Y.S, Hartiala, J., Toivonen, H., Uotila, P. (1979),
   Effects of cigarette smoke on the metabolism of vasoactive
   hormones in rat isolated lungs.
   Brit. J. Pharmacol, 65, 495.

BASKERVILLE, A. (1976).
   The development and persistance of bronchial gland
   hypertrophy and goblet cell hyperplasia in the pig after
   injection of isoprenaline.
   J. Path, 119, 35.

BAYER, B. M., Beaven, M. A. (1979),
   Evidence that indomethacin reversibly inhibits cell growth
   in the G1 phase of the cell.
   Biochem. Pharmacol., 28, 441.

BERRY, C. N., Hoult, J. R. S., Littleton, J. M., Moore, P. K.,
    Umney, N. D.
    Nicotine causes prostaglandin efflux from isolated perfused
    rat lung.
    Brit. J. Pharm., 66, 101.

CAMNER, P., Philipson, K. (1972),
    Tracheobronchial clearance in smoking discordand twins.
    Arch. Environ. Health, 25, 60.

CAMNER, P., Philipson, K., Arvidsson, T. (1973),
    Withdrawal of cigarette smoking - a study on
    tracheobronchial clearance.
    Arch. Environ. Health, 26, 90.

CHALON, J., Lowe, D.A., Orkin, L. R., (1971),
    Tracheaobronchial cytologic changes during the mentrual
    cycle.
    J. Am. Med. Ass., 218, 1928.

CHRISTENSEN, T. G., Korthy, A. L., Snider, G. L., Hayes, J. A.
    (1977).
    Irreversible bronchial goblet cell metaplasia in hamsters
    with elastase-induced emphysema.
    J. Clin. Invest, 59, 397.

COLES, S. J., Levine, L. R., Reid, L. (1979),
    Hypersecretion of mucus glycoproteins in rat airways induced
    by tobacco smoke.
    Am. J. Path, 94, 459.

COSIO, M. G., Hale, K. A., Newoehner, D. E. (1980),
    Morphologic and morphometric effects of prolonged cigarette
    smoking on the small airways.
    Am. Rev. Resp. Dis., 122, 265.

CHOO-KANG, Y. F. J., Parker, S. S., Grant, I. W. B. (1970),
    Response of asthmatics to isoprenaline and salbutamol
    aerosols administered by intermittent positive-pressure
    ventilation.
    Brit. Med. J., 4, 465.

DALHAMN, T. (1966),
    Inhibition of ciliostatic effect of cigarette smoke by
    oxalamine citrate. (3-phenyl 5B-diethylaminoehyl-1,2,4-
    oxadiazole).
    Amer. Rev. Resp. Dis., 94, 799.

DALHAMN, T. (1969),
   The anticiliostatic effect of cigarettes treated with
   oxalmine citrate.
   Amer. Rev. Resp. Dis., 99, 447.

DALHAMN, T., Rylander, R. (1971),
   Reduction of cigarette smoke ciliotoxicity by certain
   tobacco additives.
   Amer Rev. Resp. Dis., 103, 855.

EL-GHAZZAWI, I. F., Mandour, M. A., Aziz, M. T., El-Heneidy, A.
   R. (1979).
   Changes in lower respiratory epithelia induced by oestrogen
   intake.
   J. Laryngolo & Otol, 93, 601.

EL-HENEIDY, A. R., Helmy, I. D., Michael, M. A. (1966),

   Alexandria Medical Journal, 12, 275.

FERREIRA, S. H. (1972),
   Prostaglandins, aspirin-like drugs and analgesia.
   Nature (New Biol)., 240, 200.

FLOREY, H., Carelton, H. M., Wells, A. Q.
   Mucus secretion in the trachea.
   Brit. J. Exp. Path., 13, 269.

FLOWER, R. J. (1974),
   Drugs which inhibit prostaglandin biosynthesis.
   Pharmacol. Rev., 26, 33.

FLOWER, R. J. (1978),
   Steroidal anti-inflammatory drugs as inhibitors of
   phospholipase A2.
   Adv. Porstaglandin Thrombox. Res., 3, 105.

FLOWER, R. J. (1979),
   Anti-inflammatory steroids induce biosynthesis of a
   phospholipase A2 inhibitor which prevents prostaglandin
   generation. Nature, 277, 456.

GALLAGHER, J. T., Kent, P. W., Passatore, M., Phipps, R. J.,
   Richardson, P. S. (1975),
   The composition of tracheal mucus and the nervous control of
   its secretion in the cat.
   Proc. R. Soc. (Lond) B., vol. 192, p.49.

GOCO, R. V., Kress, M. B., Brantigan, O. C. (1963),
   Comparison of mucus glands in the tracheal bronchial tree of
   man and animals.
   Ann. N. Y. Acad. Sci., 106, 555.

GOODMAN, L. S., Gilman, A. (1975),
   The Pharmacological Basis of Therapeutics, 5th ed. McMillan,
   N. Y.

GREIG, N., Ayres, M., Jeffery, P. K. (1980),
   The effect of indomethacin on the response of rat bronchial
   epithelium to tobacco smoke.
   J. Path., 132, 1.

HAYASHI, M., Huber, G. L. (1977),
   Quantitative differences in goblet cells in the tracheal
   epithelium of male and female rats.
   Am. Rev. Resp. Dis., 115, 595.

HAYASHI, M., Sornberger, C.G., Huber, G. L. (1978),
   Differential response in the male and female tracheal
   epithelium following exposure to tobacco smoke.
   Chest, 73, 515.

HAYASHI, M., Sornberger, G. C., Huber, G. L. (1979),
   Morphometric analyses of tracheal gland secretion and
   hypertrophy in male and female rats after experimental
   exposure to tobacco smoke.
   Am. Rev. Resp. Dis., 119, 67.

HAYES, J. A., Christensen, T. C. (1978),
   Bronchial mucus hypersecretion induced by elastase in
   hamsters: ultrastructural appearances.
   J. Pathol., 125, 25.

HAYES, J. A., Korthy, A. L., Snider, G. L. (1975),
   The pathology of elastase-induced panacinar emphysema in
   hamsters.
   J. Pathol., 117, 1.

HOGG, J. C., Macklan, P. T., Thurlbeck, W. M. (1968),
   Site and nature of aorway obstruction in chronic obstructive
   lung disease.
   New Eng. J. Med., 278, 1355.

HUMES, J. L., Strausser, H. R. (1974),
   Prostaglandins and cyclic nucleotides in Maloney sarcoid
   tumours.
   Prostaglandins, 5, 183.

JEFFERY, P. K. (1973),
  Goblet cell increase in rat bronchial epithelium following
  irritation and drug administration: an experimental and
  electron microscopic study. Ph.D. Thesis, London
  University.

JEFFERY, P. K. (1982),
  The normal structure of bronchial epithelium.
  In: The Lung in its Environment, (ed: G. Bonsignore & G.
  Cumming), vol. 6, Life Sciences (ed: A. Zichichi), Plenum,
  N.Y. 57-72.

JEFFERY, P. K., Reid, L. (1975),
  New observations of rat airway epithelium: a quantitative
  electron microscopic study.
  J. Anat., 120, 295.

JEFFERY, P. K., Reid, L. (1977),
  The respiratory mucous membrane.
  In: Respiratory Defence Mechanisms, (Ed. J. Brain, D. F.
  Proctor and L. Reid) Monograph No. 3. of Lung Biology in
  Health and Disease.   Ed: C. Lenfant, Marcel Dekker Inc.,
  N.Y., pp. 193-245.

JEFFERY, P. K., Reid, L. (1981),
  The effect of tobacco smoke with or without
  phenylmethloxadiazole (PMO) on rat bronchial epithelium: a
  light and electron microscopic study.
  J. Path., 133, 341.

JEFFERY, P. K., Ayers, M., Rogers, D. F. (1982),
  The mechanisms and control of bronchial mucous cell
  hyperplasia.
  Chest, 815, 27S.

JEFFERY, P. K., Ayers, M., Rogers, D. F. (1982),
  The mechanisms and control of bronchial mucous cell
  hyperplasia.
  In: 2nd Symposium on Mucus in Health and Disease (ed. E.
  Chantler). Advances in Experimental Medicine and Biology,
  vol. 144, Plenum Press, pp. 399-410.

JONES, R., Bolduc, P., Reid, L. (1972),
  Protection of rat bronchial epithelium against tobacco
  smoke.
  Brit. Med. J., 2, 142.

JONES, R., Reid, L. (1978),
    Secretory cell hyperplasia and modification of intracellular
    glycoprotein in rat airways induced by short periods of
    exposure to tobacco smoke, and the effect of the anti-
    flammatory agent phenylmethyloxadiazole.
    Lab. Invest., 39, 41.

JONES, R., Reid, L. (1979).
    B agonists and secretory cell number and intra-cellular
    glycoprotein in airway epithelium. The effect of
    isoproterenol and Salbutamol.
    Am. J. Path, 95, 407.

KLEINERMAN, J., Sorensen, J., Rynbrandt, D. (1976).
    Chronic bronchitis in the cat produced by chronic
    methacoline administration.
    J. Pathol., 82, 45a.

LAMB, D., Reid, L. (1968).
    Mitotic rates, goblet cell increase and histochemical
    changes in mucus in rat bronchial epithelium during exposure
    to SO2.
    J. Path. Bact., 96, 97.

LAMB, D., Reid, L. (1969),
    Histochemical types of acidic glycoprotein produced by
    mucous cells of the trachea bronchil glands in man.
    J. Path, 98, 213.

LAMB, D., Reid, L. (1969),
    Goblet cell increase in rat bronchial epithelium after
    exposure to cigarette and cigar tobacco smoke.
    Brit. Med. J., 1, 33.

LAMB, D., Reid, L. (1970),
    Histochemical and autoradiographic investigation of the
    serous cells of the human bronchial glands.
    J. Path, 100, 127.

LEWIS, G. P., Piper, P. J. (1975),
    Inhibition of release of prostaglandins as an explanation of
    some of the actions of anti-inflammatory corticosteroids.
    Nature, 254, 308.

LUKIE, B. E., Forstner, G. C. (1972)
    Inhibition of (I-14C) glucosamine incorporation by sodium
    salicylate in vitro.
    Biochim. Biophys. Acta, 273, 380.

MAHONEY, J. M., Waterbury, L. D. (1981),
  The effect of orally administered prostaglandins on gastric
  mucus secretion in the rat.
  Prostaglandins Med., 7, 101.

MAROM, Z., Shelhamer, J. H., Kaliner, M. (1981),
  Effects of arachidonic acid, monohydroxyeicosatetranoic acid
  and prostaglandins on the release of mucus glycoproteins
  from human airways in vitro.
  J. Clin. Invest., 67, 1695.

MEYRICK, B., Reid, L. (1970),
  Ultrastructure of cells in the human bronchial submucosal
  glands.
  J. Anat, 107, 281.

MEYRICK, B, Sturgess, J., Reid, L. (1969),
  Reconstruction of the duct system and secretory tubules of
  the human bronchial submucosal glands.
  Thorax, 24, 729.

MINTY, B. D., Jordan, O., Jones, J. G. (1981),
  Rapid improvement in abnormal pulmonary epithelial
  permeability after stopping cigarettes.
  Brit. Med. J., 282, 1183.

MONCADA, S. (1979),
  Mode of action of aspirin-like drugs.
  Adv. Intern. Med., 24, 1.

MUSIL, J., Weissova, J., Adam, M., Prokopec, J. (1968),
  The influence of anti-inflammatory drugs on glycoprotein
  biosynthesis in vitro.
  Pharmacology, 1, 295.

NEDERGAARD, O. A., Schrold, J. (1977),
  The mechanism of action of nicotine on vascular adrenergic
  neuroeffector transmission.
  Euro. J. Pharmacol., 42, 315.

PHIPPS, R. J. (1981),
  The airway mucociliary system.
  In: Intern. Rev. Physiol. Respiratory Physiology, III, Vol.
  23, (ed: J. G. Widdicombe), University Pauls Press,
  Baltimore, pp. 213 - 260.

PLOPPER, C. G., Mariassy, A. T., Hill, L. H. (1980),
    Ultrastructure of the nonciliated bronchiolar epithelial
    (Clara) cell of mammalian lung: I. A comparison of rabbit,
    guinea pig, rat, hamster and mouse.
    Exp. Lung Res., 1, 139.

REID, L. (1954),
    Pathology of chronic bronchitis.
    Lancet, 1, 275.

REID, L. (1960),
    Measurement of the bronchial mucous gland layer: a
    diagnostic yardstick in chronic bronchitis.
    Thorax, 15, 132.

REID, L. (1963),
    An experimental study of hypersecretion of mucus in the
    bronchial tree.
    British J. Exp. Path, 44, 437.

REID, L. (1978),
    The cell biology of mucus secretion in the lung.
    In: The Lung Structure, Function & Disease. (ed: W. M.
    Thurlbeck & M. R. Abell), Wilkins.

REID, L. M., Jones, R. (1980),
    Mucous membrane respiratory epithelium.
    Environ. Hlth. Perspectives, 35, 113.

RICHARDSON, P. S., Phipps, R. J. (1978),
    The anatomy, physiology, pharmacology and pathology of
    tracheobronchial mucus secretion and the use of expectorant
    drugs in human disease.
    Pharmac Ther. B., 3, 441.

ROGERS, D. F., Jeffery, P. K. (1981),
    The effect of anti-inflammatory agents on the response of
    bronchial epithelium to tobacco smoke.
    Canad. Rev. Lab., Med. 3, Abstract 35.

ROYAL COLLEGE OF PHYSICIANS.
    Smoking or Health, the Third Report.
    Pitman Med. Pub. Co., London, 1977.

SPICER, S. S., Chakrin, J. W., Wardell, J. R. (1974),
    Effect of chronic sulphur dioxide inhalation on the
    carbohydrate histochemistry and histology of the canine
    respiratory tract.
    Am. Rev. Resp. Dis., 110, 13.

STURGESS, J., Reid, L. (1972),
  Secretory activity of human bronchial submucosal glands in
  vitro.
  Exp. Mol. Pathol., 16, 362.

STURGESS, J., Reid, L. (1973),
  The effect of isoprenaline and pilocarpine on (a) bronchial
  mucus-secreting tissue and (b) pancreas, salivary glands,
  heart, thymus, liver and spleen.
  Br. J. exp. Path, 54, 388.

SYKES, J. A. C., Maddox, I. J. (1972),
  Prostaglandin production by experimental tumours and the
  effects of anti-inflammatory compounds.
  Nature (New Biol), 237, 59.

VANE, J. R. (1971),
  Inhibition of prostaglandin synthesis as a mechanism of
  action for aspirin-like drugs.
  Nature (New Biol.) 231, 232.

DISCUSSION

SPEAKER: JEFFERY   **CHAIRMAN: CUMMING**

CHAIRMAN: The paper is open for discussion.

BONSIGNORE: I would like to know in which way isoprenaline
                affects ciliary activity because if mucous cell
                activity is increased and ciliary activity is
                increased as well, this implies an increase in the
                visco-elastic properties of mucus and a change in
                serous cell activity.

JEFFERY: Thank you for that question. Of course in our
                system we have no data on the functional activity
                of cilia, so I cannot say what that would be, the
                only thing I can draw on is the work of Iravani
                and Melville who showed that beta adrenergic
                stimulants would increase ciliary activity, so
                perhaps this is so in this situation. As regards
                visco-elastic properties, it is believed that a
                shift from the serous type of secretion to one
                which is more acidic means that there is an
                increase in the viscosity of mucus. Now if the
                mucus is too viscid then its coupling (which is
                specific with the tips of the cilia) is likely to
                be less effective, if it isn't viscid enough it
                also will not take place and it will not be
                shifted. The classic case of the latter is in
                patients with bronchorrhea who produce copious
                amounts of secretion which are too thin to be
                removed by the cilia.

DENISON: You mention in your studies the extension of
                mucous cells into the small airways and I would
                like to know whether that's accompanied by cilia
                there, because presumably without cilia there
                would be mucus stagnation.

JEFFERY: Yes, its an interesting question. In fact if you
                look at normal airways whether they be human or
                rat or many other species, there are cilia present
                as far distally as the respiratory bronchioles,
                the only difference is that cilia get shorter and
                shorter as you move more peripherally. In the

trachea they are normally about 6 um high, while down in the bronchioles they are about 4 um high. What the functional consequences of that may be we don't yet know. So there are plenty of ciliated cells down in the small airways, normally. I suppose the question is if there are no mucous cells in the small airways, what are all those ciliated cells doing? And I really can't answer that. But the observation with our tobacco smoke studies, and in many of the effects with drugs, is that the normal or greater than normal numbers of ciliated cells are present and morphologically they look normal. Presumably they are functional and perhaps one could look at it as a protective response in that now any mucus which begins to be produced there would be moved by the cilia. Finally I think it comes back to the question: is the mucus which begins to be produced there would be moved by the cilia. It comes back to the question: is the mucus in the right amount and is it of the right visco-elastic properties? There is something happening in the small airways with regard to the loss of Clara cells (whatever function the Clara cell may be) and the appearance of mucus at this distal site. Perhaps there may be an imbalance of surfactant in the small airway with regard to the amount of surfactant lining the alveoli and this could lead to early airways closure, which with the presence of mucus would be functionally not beneficial to either the experimental animal or the patient with small airways disease.

LEE: Something you said in that last statement about the cilia being there with or without mucous cells in the peripheral airways. If there are no cilia there, we would be in real trouble with inflammation, because with the invasion of the peripheral airways by macrophages, you would be in terrible trouble, surely.

JEFFERY: All right, there is a very good suggestion for the function of cilia in a distal airway. We produce millions of macrophages for the normal defence system in the alveoli, some of the macrophages escape from the lung via the lymphatics, as we know, but most of them come up the muco-ciliary escalator, so perhaps the function of those distal cilia is to remove macrophages laden with pollutant up the airways to the pharynx where they would normally be swallowed.

RICHARDSON: I'd like to follow up that question. It has been seen by Silberberg's group that in the large airways the cilia's only function is moving particles providing that mucus is present. Is there any functional specialisation of the cilia in the small airways which suggests that they might be different in their function, that they might be able to remove fluids and particles in the absence of mucus. Do they for instance have hooks on their ends, like you have described in the large airways?

JEFFERY: I can say that the only difference we can detect is that they are shorter, they indeed do have the hooks that we described earlier on in the trachea and main bronchi of rats, but of course we are not sure what function these hooks have. The interesting thing that one could point out is that the longer the cilia are (and now I am going a little bit broader than the mammalian system to look at the lamellibranchs, the molluscs and the protozoa) the more specifically they are designed to move water, the shorter they are, the less able they are to move water, this may have some functional consequence from the point of view of moving periciliary fluid layer which lines the respiratory tract, but I am afraid that statement goes no way to really answering the question with regard what cilia do in the distal airways.

ROSSI: In your system you use cigarette smoke to produce irritation of the mucous membrane. One mechanism by which the cigarette smoke is acting is probably by oxidant radicals. Did you use in your experiments, on mucous secretion, oxidant radical scavengers like vitamin E, for example.

JEFFERY: No, its a very good question and a very good experiment to do. In the past we have not used oxidant radical scavengers, nor have we used specific oxidants, to see whether they will have a similar effect. We are however beginning experiments with substances which are specifically designed to act as anti-oxidants and perhaps if we have another such conference, I'll be able to give you the results of our current studies in this field.

QUESTIONER: You'll commonly find that when elderly patients,
               who have been smoking for quite a long time, with
               chronic bronchitis and considerably severe
               bronchorrhea are admitted to hospital and they
               give up smoking they produce virtually no more
               sputum.   Do you think that suddenly stopping
               smoking alters the mucus structure?   And then, to
               go back to the previous subject concerning ciliary
               function, in these patients the cilia should be in
               constant movement.

JEFFERY: I think these are interesting observations, i.e.
               how long it took after the cessation of smoking
               for the sputum volume to diminish.   Clearly in our
               animal studies after only two weeks' exposure you
               can recover really rather quickly from the effects
               of tobacco smoke, structurally speaking.   And
               indeed I didn't mention the recovery of the type
               of secretion, the switch they receive from neutral
               to acidic glycoprotein with exposure to tobacco
               smoke recovers even quicker than does the total
               cell number.   So perhaps this is also happening in
               your patients but of course your patients have a
               much longer standing exposure to smoke and they
               have the added complication of recurrent
               infection.   I would think, as a result of changes
               to their bronchial tree (and this we saw with the
               isoprenaline studies) they took much longer to
               recover and this could well be due to a greater
               degree of mucosal damage.

               With regard to the ciliary studies I return to the
               study of Cameron's group, who showed that it takes
               some three months or more for tracheo-bronchial
               clearance to return to normal, after cessation of
               smoking.   Indeed the effects of tobacco smoke on
               the muco-ciliary system are by no means clear,
               there are as many who say that prolonged exposure
               to tobacco smoke will stop cilia, as others who
               say that tobacco smoke will initially stop them,
               following which time there will be a recovery and
               maintain some of ciliary function in spite of
               continued exposure to tobacco smoke.   So the
               defect in muco-ciliary clearance which one
               observes in relation to tobacco smoke could well
               be a mucous effect rather than an effect on
               ciliary motility itself.

HEATH: Could I ask a silly question with timidity and bravery? You tell us how indomethacin inhibits mucous secretion. Let us take a state of affairs where a patient is producing massively sticky sputum, what effect will indomethacin have? Because it is presumed it could be bad for the mechanism, for clearing the sputum: on the other hand it could be in fact efficacious.

JEFFERY: Let me say that I don't wish that you all go out and begin giving your patients indomethacin. Obviously much more work needs to be done and perhaps some clinical trials are envisaged in the not too distant future. One must be careful, our results only show indomethacin inhibits the increase in mucous cell number and we haven't done the in vitro work which is required to show that in every case that I have shown you today the increase in mucous cell number parallels hypersecretion, (i.e. an increase in discharge of secretion by the cells). So all that we are seeing here is the effect on the number of mucus-secreting cells; we have no idea at this stage as to the effect of the drugs on the amount of secretion which is coming out of these cells. The only drugs which have been examined in this way, interestingly enough, are those drugs, which affect the microtubules within the cells, vinoblastine, cytochalasin B and colchicine. These three drugs, in culture, are shown to inhibit the amount of mucus secreted by those cells, that was in a system using pig airway epithelium. So in answer to your question I have no idea of the effect of indomethacin on the amount secreted or in fact on the nature of the secretion in terms of visco-elastic properties.

THE ACTION OF DRUGS ON THE MECHANISMS

WHICH KEEP THE AIRWAYS CLEAR

P. S. Richardson

Department of Physiology
St. Georges Hospital Medical School
Cranmer Terrace, Tooting, London

There are three mechanisms which prevent dust, secretions and cells from accumulating in the airway lumen: coughing, the mucociliary transport system and phagocytosis by macrophages and polymorphonuclear leukocytes. Where these mechanisms fail some airways are likely to become blocked, giving rise to poor ventilation to parts of the lung. In extreme cases there is extensive bronchial and bronchiolar plugging and patients may die from ventilatory failure. This is seen most commonly in those who die from asthma: at post mortem the conducting airways contain extensive plugs of a material which consists of mucus, plasma exudate, fibrin and cellular debris (Dunnill, 1960). There is growing evidence that patients who suffer from milder asthma may also accumulate lesser quantities of such material in their conducting airways. Dunnill (1975) has shown that, in a small series of patients who died in traffic accidents while in remission from their asthma, there was partial occlusion of many airways by secretions and cellular debris.

It is worth considering the consequence of partial airways obstruction by material within the ring of bronchial smooth muscle. A minor accumulation would cause only a small increase of airway resistance in the relaxed airway, but once the smooth muscle of the airway constricted the uncleared material within the lumen would sharply exacerbate the rise in resistance (Freedman, 1972). Fig. 1 gives an example of this effect. This is likely to be one of the factors which causes bronchial hyper-reactivity, the severe increase in airways' resistance in asthmatic subjects who are exposed to a bronchoconstrictor stimulus which, in normal people, would cause only a trivial increase in airway resistance.

117

118 P. S. RICHARDSON

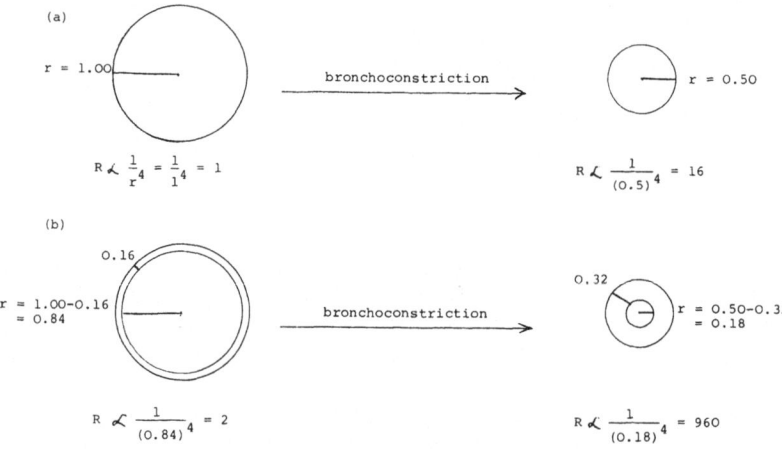

Figure 1

The effect of bronchoconstriction on the resistance of an
airway with an unimpeded lumen (a) and an airway of the same
size but with its lumen partly blocked by uncleared secretion
(b). R = resistance; r = radius of the airway lumen.   The
calculation of airway resistance, in arbitrary units, assumes
that Poiseuille's law applied (R $\propto \frac{1}{r^4}$) It can be seen that a

smear of uncleared secretion which doubles airway resistance in
the relaxed airway increases it profoundly in the constricted
airway.

The remainder of this paper considers the effect of drugs on bronchial clearance by mucociliary transport, probably the most important of the clearance mechanisms considered in the first paragraph.

Action of drugs on mucociliary clearance

The cilia on the epithelial lining of the airway beat rhythmically so that their tips engage an overlying sheet of mucus in their forward thrust but disengage during the recovery stroke. The cilia are thought to beat in a watery layer of fluid (the periciliary fluid) and the mucus to lie on top of this. The presence of a layer of mucus, or something with similar physical properties (King, 1979), is essential for effective ciliary beating: in its absence the beating continues but fails to move particles placed on the ciliated surface. There are a number of variables on which drugs might act to change the effectiveness of mucociliary transport and Table 1 summarises these. Unfortunately some of these cannot be measured with any precision (viz. the force of ciliary beating, the co-ordination of cilia and the rate of secretion of periciliary fluid) so the account which follows is necessarily incomplete. One important thing which can be estimated is the rate of movement of particles on a ciliated airway surface. In man there are two main methods which have been used for this; the movement, judged by x-ray pictures or by α-camera pictures, of relatively large particles placed in the trachea via a bronchoscope, and the clearance of radioactive dusts, inhaled into the airways, from the lung fields. These methods are reviewed by Pavia et al., 1980. The former method gives precise information about mucociliary transport rate in a single airway which may not be representative of the intrapulmonary airways. The latter method gives less precise information about the speed of mucociliary transport but does allow comparison of the rate of removal of dust from the entire lung before and after drug treatment. The chief disadvantage is that a change in the state of the lungs (an alteration in bronchial calibre produced by a drug under investigation, for instance) may bias the deposition of the radioactive dust into larger airways (which, being nearer the "finishing line", can be cleared faster) or smaller airways (cleared more slowly). Ideally information from both methods should be available to allow one to judge the effect of a drug on mucociliary transport.

TABLE 1

Factors affecting the rate of mucociliary transport

Ciliary beating: force

                  frequency

                  co-ordination

Mucus secretion: quantity

                  physical properties (viscosity,

                  elasticity, surface stickiness)

Secretion of periciliary fluid

Drugs affecting cholinergic receptors

Camner et al. (1974) have shown that injection of a muscarinic receptor agonist increases the rate of clearance of radioactive dust from the lungs of human subjects. Treatment with atropine slows mucociliary clearance in man (Yeates et al., 1975), suggesting that there is parasympathetic tone to some part of the mucociliary clearance system. In view of this it is surprising that inhalations of ipratropium bromide at doses substantially greater than that required to relax bronchial smooth muscle should leave mucociliary clearance rates unaffected (Francis et al., 1977).

How do these drugs act? Cholinomimetic drugs and parasympathetic nerves are known to increase mucus secretion from airways, an effect which atropine blocks (Florey et al., 1932). It is possible that an increase in the volume of mucus secreted could itself increase the rate of removal of dust by providing a more complete covering of mucus to the ciliated cells. In addition King & Viires (1979) has described how cholinergic drugs change the quality of the mucus secreted. Low doses of methacholine reduced the viscoelastic properties of mucus collected from the dog trachea and a similar change might lead to the more rapid mucociliary transport seen in the human lung by Camner et al. (1974). At higher doses methacholine produced a mucus of greater viscosity and elasticity than the control secretion and King & Viires (1979) showed that a ciliated surface transported this more slowly. The explanation for these rather complicated changes in the physical properties of mucus with increasing doses of cholinergic agonist may be that they control two relatively independent processes; water secretion and mucus glycoprotein secretion (Basbaum et al, 1981). If the former has a lower threshold than the latter, it could account for these results.

The action of cholinergic drugs on ciliary beating and the secretion of periciliary fluid are uncertain. Corssen & Allen (1959) showed that large doses of acetylcholine would increase ciliary beat frequency, but it is possible that this effect was secondary to the changes in mucus secretion described above. The action on periciliary fluid production is even less certain: Marin et al (1976) showed that cholinergic drugs stimulate the secretion of salt and hence water into the airway lumen, but whether this forms periciliary fluid, part of the mucus layer,or both, is unknown.

The disadvantage of using cholinergic drugs in the hope that they will assist in the removal of secretions and debris from

the airway is clear: they would also cause a
bronchoconstriction. It is therefore unlikely that cholinergic
drugs can ever be used in the therapy of respiratory disease.

Drugs affecting adrenoceptors

This account will ignore alphabetical convention because so
little is known about the effects of β-adrenoceptor
stimulation on mucociliary transport. Most groups which have
tested the effects of β-adrenoceptor stimulating drugs have
found that they hasten mucociliary clearance, but Bateman et al
(1979) failed to confirm this. Pavia et al (1980) have
recently reviewed this vexed area of investigation. A likely
explanation for the disagreements is that β-adrenoceptor
agonists act on mucociliary transport only in the trachea and
largest bronchi. Methods which measure mucus transport
predominantly in these airways will demonstrate an effect on
mucociliary transport while those that examine whole lung
clearance of particles will show little change in transport
rate. All the drugs whose effects on mucociliary transport
have been tested so far are either mixedβ_1- and β_2-receptor
agonists or selective β_2-receptor agonists. Thus there is no
information on whether α- and β_1-receptors can affect
mucociliary clearance.

Drugs which stimulate β-adrenoceptors may act by augmenting
mucus production. Such an effect had been shown in both cat
(Peatfield & Richardson, 1982) and man (Phipps et al, 1982).
Drugs affecting α-adrenoceptors also provoke mucus secretion in
cat (Phipps et al, 1980) and man (Shelhamer et al, 1980;
Phipps et al, 1982). The only information on the change in the
quality of mucus secreted in response to adrenoceptor
stimulation is indirect: Basbaum et al (1981) showed that
adrenoceptor stimulating agents caused an emptying of
glycoprotein granules from serous cells of the ferret trachea.
This effect was distinct from that of cholinergic agonists
which caused cell vacuolation as well as glycoprotein
discharge. The authors suggested that vacuolation occurs with
salt and water discharge. This would imply that adrenoceptor
agonists which cause little vacuolation would produce a viscid
secretion, poor in water.

There is some evidence that β-adrenoceptor agonists may
increase the rate of ciliary beating but, as with cholinergic
drugs, this could be secondary to changes in the quality and
quantity of mucus secreted. Phipps et al (1980) found that
both α- andβ_2-adrenoceptor stimulating drugs increased salt and
water secretion into the airway, but again it is uncertain
whether this was part of mucus secretion or an increase in the
formation of periciliary fluid.

Other drugs

The action of a number of expectorant drugs on the rate of mucociliary clearance from the human lungs has been tested and are reviewed by Pavia et al (1980) and Richardson & Phipps (1978). These include bromhexine, glyceryl guaiacolate, S-carboxymethyl cysteine and 2-mercapto-ethane sulphonate. The results are contradictory but at least suggest that some of these expectorant agents hasten mucociliary transport. The studies give no clue as to their mechanism of action.

Pavia et al (1978) reported that subjects who inhaled an aerosol of hypertonic saline had a significantly faster clearance of radioactive particles from their lung fields than normal. Such inhalations are irritant and so may produce reflex increases in airway mucus secretion, probably by a cholinergic reflex (Phipps & Richardson, 1976). It would be interesting to see whether atropine blocked this effect.

The action of drugs on mucociliary clearance in asthmatic patients

The introduction of this paper pointed to a failure of mucociliary clearance in asthmatic patients. There is good evidence that clearance of radioactive particles from the lung fields of such patients is slow (Pavia et al, 1980) both for asthmatics in remission and those during asthma attacks. Mezey et al (1978) found that the movement of radio-opaque particles in the trachea of asthmatics in remission was only about half than in normal subjects. Subsequent antigen challenge slowed mucociliary to only a quarter of that in control subjects. There are now some important pharmacological clues which may help to explain the slowing of mucociliary transport in asthmatics. Mezey et al (1978) found that treatment of their asthmatic subjects with sodium cromoglycate before antigen challenge prevented the slowing of mucociliary transport. Subsequently the same group found that treatment of asthmatic subjects with FPL 55712, an antagonist of Slow Reacting Substance of Anaphylaxis (SRS-A or leukotrienes), also prevented the slowing of mucociliary transport on antigen challenge (Ahmed et al, 1980). Indeed tracheal mucous velocity increased on challenge in those protected with FPL 55712. Their tentative explanation of these findings is that challenge of sensitised lungs with allergen provokes the release of a number of mediators including histamine, prostaglandins and SRS-A. Of these only the last inhibits mucociliary transport while the others between them increase transport rate. There is some recent work which shows that SRS-A increases the release of mucus glycoproteins from human airway in vitro and

that FPL 55712 inhibits this action (Shelhamer et al 1982).
Most drugs that increase mucus secretion also accelerate
mucociliary transport (see above), but it is possible that SRS-
A may cause the secretion of mucus with abnormal
viscoelasticity, outside the range where it is readily
transported by cilia. In view of the apparently viscid nature
of the sputum of asthmatics it is likely that the airway mucus
becomes too highly viscoelastic but, surprisingly, this has
never been adequately tested. Alternatively SRS-A might have a
detrimental effect on cilia or, by increasing capillary
permeability, flood the airway with plasma exudate.

Conclusions

This review began by pointing to the consequences of
uncleared secretions in the airways and proceeded to describe
the drugs which change mucociliary transport rate, but an
acceleration of mucociliary transport rate does not necessarily
prevent airway blockage because this is also influenced by
changes in secretory rate. An ideal drug would diminish airway
secretion and hasten mucociliary transport (and other forms of
airway clearance). The worst type of agent would increase
secretory rate and prevent mucociliary transport altogether
(SRS-A may act in this way). Most of the drugs and mediators
discussed here increase both mucociliary transport and
secretory rate. This will only be an advantage if the former
increases by more than the latter. Unfortunately we still lack
any adequate means of testing whether this is so in patients.

REFERENCES

Ahmed,T., Greenblatt,D.W., Birch,S., Marchette,B. and Wanner,A.
    (1981).
    Abnormal mucociliary transport in allergic patients with
    antigen induced bronchospasm: role of Slow Reacting
    Substance of Anaphylaxis.
    Am. Rev. resp. Dis., 124, 110-114.

Basbaum,C.B., Ueki,I., Brezina,L. and Nadel,J.A. (1981).
    Tracheal submucosal gland serous cells stimulated
    in vitro with adrenergic and cholinergic agonists.
    Cell and Tissue Research, 220, 481-498.

Bateman,J.R.M., Lennard-Jones,A.M., Pavia,D. and Clarke,S.W. (1980).
      Lung mucociliary clearance of normal subjects during selective and non-selective β-blockade.
      Clin Sci. 58, 6P.

Camner,P., Strandberg,K. and Philipson,K. (1974).
      Increased mucociliary transport by cholinergic stimulation.
      Arch environm. Health, 29, 220-224.

Camner,P., Standberg,K. and Philipson,K. (1976).
      Increased mucociliary transport by adrenergic stimulation.
      Arch. environm. Health, 31, 79-82.

Corssen,G. and Allen,G.R. (1959).
      Acetylcholine: its significance in controlling ciliary activity of human respiratory epithelium in vitro.
      J. appl. Physiol., 14, 901-904.

Dunnill,M.S. (1960).
      The pathology of asthma with special reference to changes in the bronchial mucosa.
      J. clin. Path. 13, 27-33.

Dunnill,M.S. (1975).
      Morphology of the airways in asthma.
      pp 213-221 in New Directions in Asthma. Ed Myron Stein.
      Publ. Am. Coll. Chest Physicians. Park Ridge, Illinois.

Florey,H., Carleton,H.M. and Wells,A.Q. (1932).
      Mucus secretion in the trachea.
      Br. J. Exp. Path. 13, 269-284.

Francis,R.A., Thomson,M.L., Pavia,D. and Douglas,R.B. (1977).
      Ipratropium bromide: mucociliary clearance rate and airways resistance in normal subjects.
      Brit. J. Dis. Chest. 71, 173-178.

Freedman,B.J. (1972).
      The functional geometry of the bronchi.
      Bull. Physiopathol. resp. 8, 545-551.

King,M. (1979).
      Interrelation between mechanical properties of mucus and mucociliary transport: effect of pharmacologic interventions.
      Biorheology, 16, 57-68.

King,M. and Viires,N. (1979).
    Effect of methacholine chloride on rheology and transport
    of canine tracheal mucus.
    J. appl. Physiol., 47, 26-31.

Marin,M.G., Davis,B. and Nadel,J.A. (1976).
    Effect of acetylcholine on Cl^- and Na^+ fluxes across dog
    tracheal epithelium in vitro.
    Am. J. Physiol., 231, 1546-1549.

Mezey,R.J., Cohn,M.A., Fernandez,R.J., Jannokiewicz,A.J. and
    Wanner,A. (1978).
    Mucociliary clearance in patients with antigen-induced
    bronchospasm.
    Am. Rev. resp. Dis. 118, 677-684.

Pavia,D., Bateman,J.R.M. and Clarke,S.W. (1980).
    Deposition and clearance of inhaled particles.
    Bull. europ. Physiopathol. resp. 16, 335-366.

Pavia,D., Thomson,M.L. and Clarke,S.W. (1978).
    Enhanced clearance of secretions from the human lung
    after the administration of hypertonic saline aerosol.
    Am. Rev. resp. Dis. 117, 199-203.

Peatfield,A.C. and Richardson,P.S. (1982).
    The control of mucin secretion into the lumen of the cat
    trachea by - and -adrenoceptors and their relative
    involvement during sympathetic nerve stimulation.
    Eur. J. Pharmacol. 81, 617-626.

Phipps,R.J., Nadel,J.A. and Davis,B. (1980).
    Effect of -adrenergic stimulation on mucus secretion and
    on ion transport in cat trachea in vitro.
    Am. Rev. resp. Dis. 121, 359-365.

Phipps,R.J. and Richardson,P.S. (1976).
    The effect of irritation at various levels of the airway
    upon tracheal mucus secretion in the cat.
    J. Physiol., 261, 563-581.

Phipps,R.J., Williams,I.P., Richardson,P.S., Pell,J., Pack,R.J.
    and Wright,N. (1982).
    Sympathomimetic drugs stimulate mucin secretion from
    human bronchi.
    Clin. Sci. 63, 23-28.

Richardson,P.S. and Phipps,R.J. (1978).
   The anatomy, physiology, pharmacology and pathology of
   tracheobronchial mucus secretion and the use of
   expectorant drugs in human lung disease.
   Pharmacol. Ther. B., 3, 441-479.

Shelhamer,J.H., Marom,Z. and Kaliner,M. (1980).
   Immunologic and neuropharmacologic stimulation of mucous
   glycoprotein release from human airway in vitro.
   J. Clin. Invest., 66, 1400-1408.

Yeates,D.B., Aspin,N., Levison,H., Jones,M.T. and Bryan,A.C.
   (1975).
   Mucociliary tracheal transport rates in man.
   J. Appl. Physiol., 39, 487-495.

DISCUSSION

SPEAKER: RICHARDSON   **CHAIRMAN: G. CUMMING**

LEWIS:

If you are suggesting that it is an advantage to increase the rate of mucus clearance in asthmatics, it is obviously important not to give a drug which will also increase the rate of mucus production. Beta-adrenoceptor agonists increase the rate of mucous transport, but also augment its secretion. On balance are they counter-productive?

RICHARDSON:

It is a commonplace observation that the mucus produced in the lungs of asthmatics during an attack is highly viscous and elastic so its physical properties probably slow mucociliary transport (King et al., 1974). It is possible that beta-adrenoceptor agonists cause secretion of mucus with physical properties which are more favourable for mucociliary transport. If this is true, then the hastening of mucociliary transport is likely to outweigh the additional burden for the cilia to move. We are still too ignorant to say with certainty whether a drug like terbutaline will clear the airways or block them further.

DENISON:

I want to ask two questions about the interpretation of the mechanisms for mucus clearance you describe. First, if we picture the path of a particle cleared from an alveolus to the larynx it must, on average, walk through 23 generations of airways. If we put a radioactive particle into each alveolus and set it to march towards the larynx, taking steps of one generation of airway at a time, then to start with the radioactivity will be spread diffusely over the lung fields. When will a α-camera image first show any evidence that the particles have moved? I suspect that only when the particles are in the final 5 generations of airways, the largest bronchi, will any change be evident.

RICHARDSON: Alveolar clearance is a slow process with a half-life of many tens of days, while bronchial clearance has a half-life of only 4 hours (Clarke & Pavia, 1980). My talk referred only to mucociliary transport.

DENISON: You misunderstood my question. If the particles take one minute to travel from generation 23 of the airways to generation 22 and another minute to reach generation 21 and so on, you will not detect any motion until the last five minutes when they are in the largest bronchi. It is simply a question of the geometry of the airways.

RICHARDSON: Yes, I accept that something like that is probably true, though your scheme makes large assumptions about the velocity of particle transport in the small airways.

DENISON: My next question concerns how useful it is to increase the rate of mucociliary clearance. The rapid convergence of secretions from huge numbers of small airways to a few central bronchi will increase the depth of the mucous layer in the latter. As you pointed out in your talk this will block the transmission of air, the main function of the airways. Has anyone measured how airway function changes when mucociliary transport rate increases?

RICHARDSON: No, no one has. Kilburn (1968) was the first to point out the problem posed by the layers of mucus on the surface of each of the peripheral airways as they converged on the central airways. Perhaps the danger is not as great as he supposed because now it's thought that mucus travels as discrete flakes or rafts in the small airways and only forms a continuous layer in the larger bronchi. Another factor which must mitigate the problem is the slower velocity of mucociliary transport in the lung periphery. (Asmundsson & Kilburn, 1970).

CUMMING: I would like to extend this discussion of the problem of the effect of airway geometry on mucus secretion. If you assume that mucus is produced as far down as the terminal bronchioles, which is a reasonable assumption, then since there are 30,000 terminal bronchioles and their actual surface is approximately, giveing a conservative estimate, 1000 times greater than the lateral surface of the trachea, we know that the linear transport rate in the trachea is about 25 millimeters per minute. There are three possible explanations how this might be reconciled. You mentioned two of them, the first is that the rate of transport in the terminal bronchioles is slower, but it must be 1000 times slower, which seems to be a rather unrealistic rate of decrement of speed, the second possibility is that the rafts of mucus is incomplete in the small airways and becomes more complete in the large and therefore the mucus in the small airways must cover 1/1000th of the part of the surface and that seems to me not to be a very good principle. A third possibility exists and that is that the mucus as it travels up the airways is modified in its structure, by the absorption of water there is a dynamic equilibrium between the sol layer and the gel layer and the combination of all these three things may result in the conversions without undue blockage. I am not at all clear about the importance of each of these mechanisms, I wonder if you are.

RICHARDSON: I wish I were. There is another possible way to explain why the convergence of mucus does not block the central airways: that is to postulate that mucus is only produced in the peripheral airways in response to need. If a particle of dust, when it settles in a terminal bronchiole, can somehow summon the mucus raft necessary to move it, we would only need very sparse production of mucus in the lung periphery. Recently Tony Peatfield and I have been doing experiments in which we have deposited dust in the trachea of a cat and measured its effect on local mucus output. As soon as the particles touch the tracheal surface the rate of secretion increases. If such an effect is present in the bronchioles we need not postulate a resting secretion and the problem of secretions converging on the upper airways would disappear unless we breathed very dusty air.

SKIDMORE: One question and one comment. Do you have any information about the action of methylxanthines or theophylline on mucociliary transport? I would also like to point out that hypertonic saline aerosols induce bronchoconstriction in asthmatics.

Even if they increase mucociliary transport in normal subjects, they must be used very carefully indeed in asthmatics.

RICHARDSON: Thank you for the comment. I had not intended to recommend that hypertonic saline aerosols should be used in the treatment of asthma, so it is an important point to emphasize. As to your question, aminophylline does increase the rate of mucociliary transport, at least in the larger airways (Sutton, Pavia, Bateman and Clarke, 1981; Serafini, Wanner and Michaelson, 1976).

References

1. Asmundsson, T. & Kilbur, K. H. (1970). Am. Rev. resp. Dis., 102, 388-397.

2. Clarke, S. W. & Pavia, D. (1980), Br. J. Clin. Pharmacol., 9, 537-546.

3. Kilburn, H. K. (1968). Am. Rev. resp. Dis., 98, 449-463.

4. King, M., Gilboa, A., Meyer, F. A. & Silberberg, A. (1974), Am. Rev. resp. Dis., 110, 740-745.

5. Serafini, S. M., Wanner, A. & Michaelson, E. D., (1976), Bull. Eur. Pathophysiol. resp., 12, 415-422.

6. Sutton, P. P., Pavia, D., Bateman, J. R. M. & Clarke, S. W. (1981), Chest, 80, 889-892.

MULTI-DISCIPLINARY TREATMENT FOR LUNG CANCER:

PROSPECTS FOR INCREASED CURE RATES IN THE NEXT DECADE

L. A. Price & Bridget T. Hill

118 Harley Street London   and

Imperial Cancer Research Fund, London

Introduction

The overall cure rate in lung cancers has not improved·for several decades in spite of the most expert surgery and radiation therapy (1,2). This is because most patients with apparently localised tumours have disseminated micrometastases at presentation. These metastases cannot be detected by current diagnostic techniques. The relationship between our ability to detect tumours in relation to their biological development and the death of the host is illustrated diagramatically in Figure 1.

These facts indicate that lung cancers should be regarded as systemic diseases at presentation in the vast majority of patients. It follows, therefore, that any attempt to increase survival in this group of diseases over the next ten years must involve the incorporation of systemic therapy (i.,e. anti-tumour drugs) as part of the initial treatment at the time of diagnosis in combination with surgery and/or radiation therapy.

Current attitudes towards the use of chemotherapy in the treatment of lung cancer

Until very recently chemotherapy was always reserved as a treatment of last resort in patients who had relapsed following initial "primary" treatment with surgery and/or radiation. This graded therapeutic response in relation to the stage of the tumour is illustrated in Figure 2.

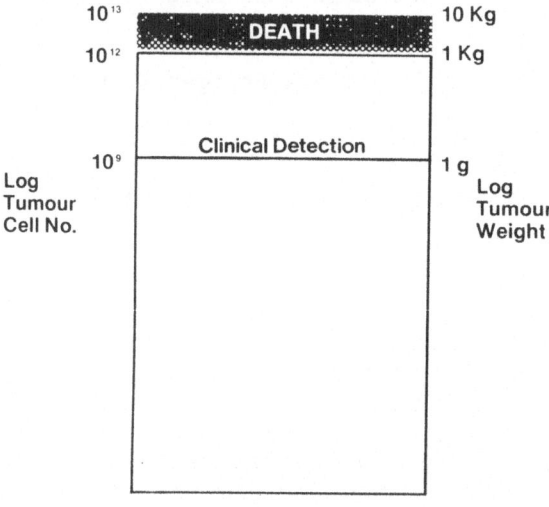

FIGURE 1

The relationship between the number of population doublings and
the increase in cell number and rate during the tumours
development, clinical detection of the tumour, and the death of
the patient. Current methods of investigation allow the
detection of the tumour only when about one gramme of tumour is
present and the tumour is already at least two thirds of the
way through its development. Therefore estimations of tumour
growth based on clinically detectable "lumps" can only be made
over a short period near the end of the tumour life span.
(After Hill and Price 1977).

TRADITIONAL APPROACH TO SOLID TUMOUR THERAPY

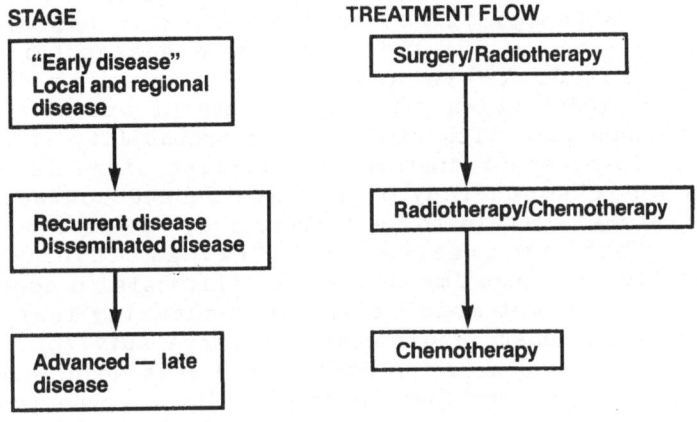

FIGURE 2

Adapted from Carter and Soper 1974

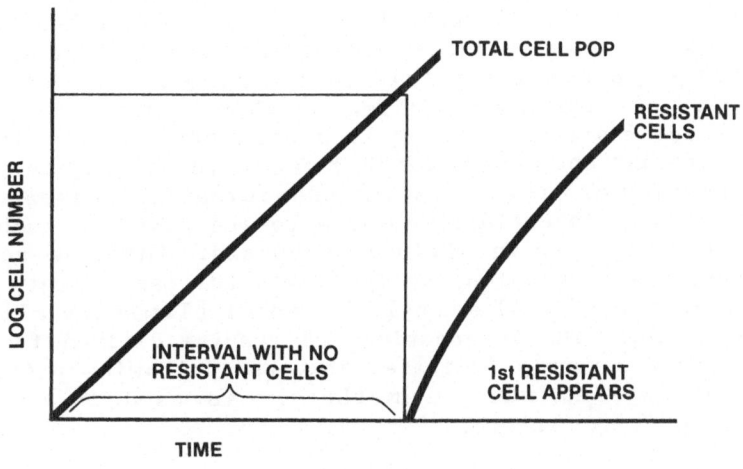

FIGURE 3

Reproduced with permission of Dr. J. H. Goldie

There are three reasons why this traditional approach should be abandoned. First, as already mentioned, local therapy cannot be expected to cure patients with tumours most of which are disseminated at the time of presentation. Second, the therapeutic strategy illustrated in Figure 2 has failed to improve the cure rate in this group of diseases for several decades. Third, there has recently been described a mathematical model which relates the rate of mutation to drug resistant phenotypes with time, to the probability of cure (3). This model is of major therapeutic interest since it not only explains why chemotherapy given late in the course of lung cancer is almost certain to be ineffective, but also because it suggests that an alternative strategy involving the administration of drugs immediately at diagnosis in combination with other treatment modalities might actually increase the cure rate or at least significantly improve survival times in patients with lung cancer. Stated simply this model postulates that the number of mutations to drug resistant phenotypes in a tumour will increase with time. This concept is illustrated in Figure 3.

Thus at the beginning of the biological life of a tumour there is a period during which there are no drug resistant phenotypes and during this time, by definition, drug therapy is very likely to be effective and even curative. However, as the tumour becomes biologically older there will occur, by random chance, mutations to drug resistant phenotypes. Therefore the "older" the tumour the more likely there is to be a high proportion of drug resistant phenotypes and consequently a significantly reduced likelihood of cure with chemotherapy. This concept may be illustrated by plotting the probability of the presence of drug resistant phenotypes against time as shown in Figure 4. Essentially this is the same as a plot of the probability of cure against time. It shows clearly that early in the life of a tumour there is a high probability of cure using drugs but the chance of this occurring falls sharply as the proportion of drug resistant phenotypes increases. This transition from potentially curable to incurable is much more sudden than might be intuitively supposed. This theoretical study supports conclusions derived over the past twenty years from experimental, (4) animal, (5) and preliminary clinical studies (6,7,8). The information and results derived from all these sources strongly indicates that the following principles should be applied when chemotherapy is used in the drug treatment of lung cancers:

1)   There is a better likelihood of curing tumours early in their growth history than when they have reached an advanced stage.

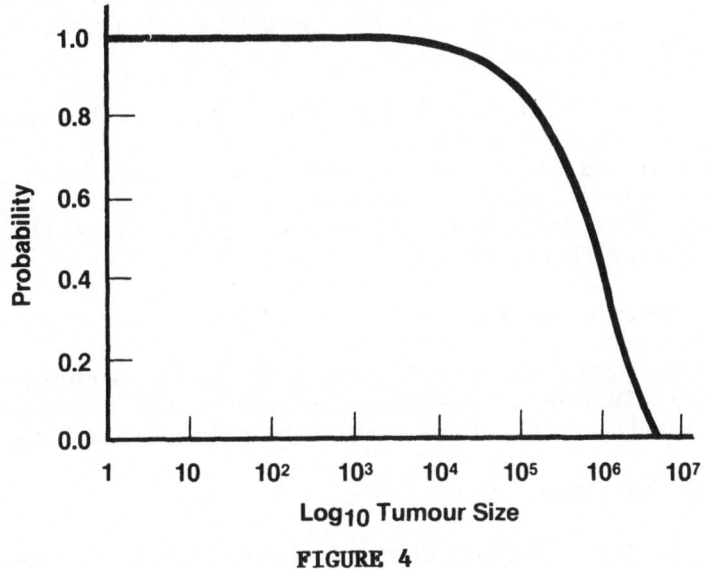

FIGURE 4

Reproduced with permission from Dr. J. H. Goldie.

2)   Combination chemotherapy using two separate groups of drugs which are effective and independent in their mode of action should reduce the spontaneous mutation rate to resistance of the tumour if the groups are given alternately (i.e. A,B,A,B etc.)

3)   Adjuvant chemotherapy should be initiated simultaneously with or even before the attack on the primary tumour with surgery and/or radiation since at a critical point in the development of the tumour expectation for cure falls rapidly and delay of even a few days will greatly reduce the likelihood of cure.

4)   Alternative non-cross resistant chemotherapy combinations are preferable to sequential chemotherapy in the prevention of the emergence of doubly-resistant mutant tumour cells.

The clear implication of these experimental and theoretical findings is that chemotherapy should be used as part of the initial combined attack on lung cancer at presentation and that the traditional role of chemotherapy depicted in Figure 2, should be abandoned. Our proposed new approach is illustrated diagramtically in Figure 5.

Clinical Considerations

From the point of view of drug therapy it is convenient to divide the histologic sub-groups of lung cancer into two main types: small cell carcinomas and non-small cell carcinomas.

Small Cell Carcinoma

This histologic sub-type of lung cancer is extremely sensitive to chemotherapy with a variety of agents. Table I lists drugs which are effective in this condition.

VP16-213 (Vepesid) is the most effective single drug against this tumour and a valuable component of combination chemotherapy which is almost certainly superior to single agent treatment in this disease. Response rates to chemotherapy of the order of 70% can be obtained, and in those patients who achieve a complete remission following two treatments (i.e. within approximately six or eight weeks) a very significant increase in survival can be achieved. For example, in "limited" disease (i.e. tumour confined to one hemi-thorax with or without involvement of supraclavicular nodes) who are treated with initial chemotherapy and achieve a complete

LOGICAL APPROACH TO SOLID TUMOUR CHEMOTHERAPY

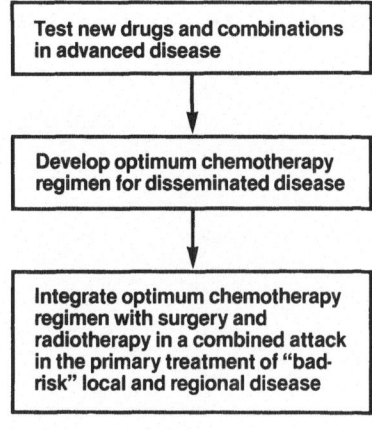

FIGURE 5

Adapted from Carter and Soper 1974

**Kinetic Classification of Antitumour
Drugs given over 24 hours against Normal
Bone Marrow stem cells**

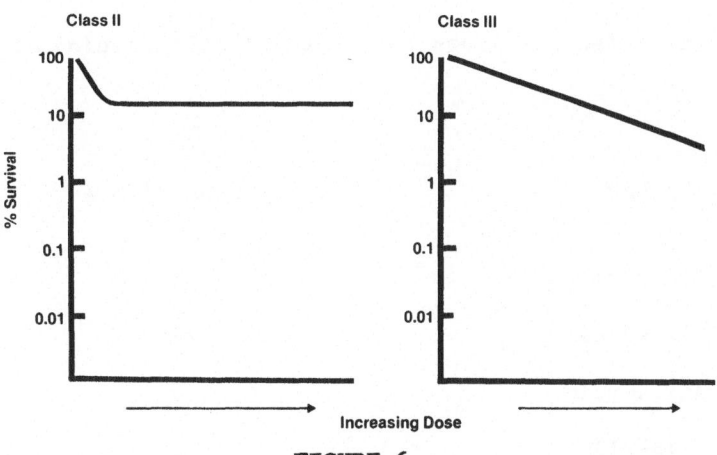

FIGURE 6

Adapted from Bruce et al 1966

TABLE I

Some effective drugs in small cell lung cancers

DRUG	RESPONSE RATE %
VP16-213	45
Vincristine	42
Nitrogen mustard	44
Adriamycin	31
Methotrexate	30
Cyclophosphamide	28
Procarbazine	28

TABLE II

Some effective drugs in non-small cell carcinomas

DRUG	RESPONSE RATE %
Mitomycin C	27
Vindesine	23
Adriamycin	18
VP16-213	18

response live significantly longer than those treated with radiotherapy or surgery along (9).

Non-small Cell Carcinoma

This group of tumours is significantly less responsive to chemotherapy. A list of drugs which have shown some effect in these tumours is provided in Table II.

At the time of writing response rates of between 30 and 40% can be achieved in patients with disseminated non-small cell carcinomas using combinations containing at least two of these drugs (10). However, there is often no associated significant increase in overall survival associated with these responses. The potential role of these combinations therefore is most likely to be in combination with surgery or radiation at the time of diagnosis. Randomised prospective controlled clinical trials are of course necessary to establish whether or not this new approach has any definite value. Since the necessary conditions for improving the treatment of lung cancer already exist, as outlined above, and since randomised prospective controlled clinical trials are necessary to determine whether these necessary conditions are sufficient conditions, it is a cause for concern that less than 1% of patients with lung cancer in the United Kingdom are entered into such trials. The main reasons why such trials are not undertaken appears to be that most physicians feel that intensive chemotherapy is automatically associated with unacceptable side effects. This belief is completely unjustified. We have clearly shown over the past 12 years that there now exists an experimentally based logical method of giving cancer chemotherapy intensively without loss of therapeutic effect and far more safely than in the past (11,12,13). The validity of this approach has now been confirmed by several other workers (14).

The theoretical basis of this approach is that anti-cancer drugs given over 24 hours can be divided into two completely separate groups according to their effects on bone marrow and tumour stem cells (5). The two groups are shown in Figure 6.

Drugs in Class II can be given in full doses over 24 hours with maximum selective toxicity against tumour stem cells but without increasing damage to normal bone marrow stem cells. Drugs in Class III however, can also achieve maximum selectivity against tumour stem cells when given over 24 hours, but are also additively toxic to the bone marrow. Thus in combination chemotherapy given over 24 hours full doses of drugs in Class II may be given but a proportionate reduction of the drugs in Class III is necessary. Drugs which fall into each group and which are widely used in the treatment of lung cancer are listed in Table III.

These experimental kinetic studies have indicated how cancer chemotherapy can be given much more safely clinically. Drug combinations should be given:

1)   over 24 - 36 hours.

2)   with proportionate reduction of Class III agents.

3)   in intermittent courses after marrow recovery

4)   as early as possible in the course of the disease, i.e. before or immediately after local therapy.

The clinical relevance of this approach has been proved in breast cancer, head and neck cancer, non squamous cell lung cancer, testicular teratomas and lymphomas (14).

The advantages of this kinetically based approach for adjuvant chemotherapy are:

1)   Full dose intensive combination chemotherapy can be given early and safely.

2)   Intervals between courses can be the minimum consistent with clinical tolerance especially for the first four or five courses since there is no severe myelosuppression.

3)   These chemotherapy protocols can be alternated using non cross resistant combinations and also integrated successfully and safely with surgery and/or radiation therapy.

4)   These are the necessary requirements for increased patient survival.

Conclusions

There now exists therefore a conceptual and practical basis for a major attempt to improve the hitherto gloomy prognosis for lung cancer patients. There is a major ethical requirement for physicians treating these diseases to take part in randomised prospective controlled clinical studies comparing chemotherapy, given according to the principles outlined above, combined with standard local treatment versus local treatment alone. Regretably the number of patients with pulmonary

TABLE III

**Kinetic classification of drugs commonly used
in the treatment of lung cancer**

CLASS II   CLASS III

Vincristine   Adriamycin

Methotrexate   Nitrogen mustard

VP16-213   Cyclophosphamide

Procarbazine   Mitomycin C

                                  5-Fluorouracil

See Text. Bleomycin is not included in this classification.

For a clarification of the various names given to these two

classes of drugs see Price and Hill 1974.

tumours in the United Kingdom entered into such trials is negligible (approximately 1%) (15). This means that one of the commonest and most lethal respiratory diseases is being virtually ignored at a time when there exists a significant probability of improving the quality of life and survival for patients with this group of tumours (16 & 17).

REFERENCES

1. Silverberg, E. 1981.
      Cancer Statistics, American Cancer Society Professional
      Education Publications.

2. Cancer Statistics Surviva. 1980.
      Office of Population Cencuses and Surveys.   Series
      M.B.I. No. 3. London.
      Her Majesty's Stationery Office.

3. Goldie, J. H. and Coldman, A. J. (1979).
      A mathematical model for relating the drug sensitivity
      of tumours to their spontaneous mutation rate.
      Cancer Treat. Rep., 63, 1727.

4. Skipper, H. E., Schabel, F. M.Jr., & Wilcox, W. S. 1964.
      Experimental evaluation of potential anticancer agents.
      XIII. On the criterial and kinetics associated with
      the "curability" of experimental leukaemia.
      Cancer Chemother. Rep., 35, 1.

5. Bruce, W. R., Meeker, B. E. and Valeriote, F. A. 1966.
      Comparison of the sensitivity of normal haematopoetic
      and transplanted lymphoma colony-forming cells to
      chemo-therapeutic agents administered in vivo.
      J. Natl. Cancer Inst. 37, 233.

6. Cooper, R. G., Holland, J. F. & Glidewell, O. 1979.
      Adjuvant chermotherapy of breast cancer.
      Cancer, 44, 793.

7. Bonadonna, G. and Valagussa, P. 1981.
      Dose-response effect of adjuvant chemotherapy in breast
      cancer 1980.
      New Eng. J. Med., 304, 10.

8. Price, L. A. and Hill, Bridget T. 1982.
        Safe and effective 24 hour combination chemotherapy
        without cis-platin as initial treatment in Head and
        Neck Cancer: Improved Survival at 5 years.
        Proceedings of Eighteenth Annual Meeting of the
        American Society of Clinical Oncology, St. Louis, C-
        786, p 202.

9. Livingston, R. B. and Greenstreet, R. L. 1982.
        Reinduction prolongs survival in complete responders
        with small cell lung cancer.
        Proceedings of the Eighteenth Annual Meeting of the
        American Society of Clinical Oncology, St. Louis. C588,
        p.151.

10. Weick, J. K., Joseph, D., Purvis, M. D. & Livingston, R.B.
        1982.
        Divided dose Mitomycin C and Vinblastine in Extensive
        Non-small Cell Lung Carcinoma.
        Proceedings Eighteenth Annual Meeting of the American
        Society of Clinical Oncology, St. Louis. C588, p.151.

11. Price, L. A. 1973.
        A kinetic approach to the drug treatment of solid
        tumours.
        In: The Third Vinca Alkaloids Symposium Ed. Shedden.

12. Price, L. A. & Goldie, J.H. 1971.
        Twenty-four hour combination chemotherapy for
        disseminated malignant disease.
        Brit. Med. J., 4, 336.

13. Hill, Bridget T. & Price, L.A. 1982.
        An experimental biological basis for increaseing the
        therapeutic index of clinical cancer chemotherapy.
        Annals of the New York Academy of Sciences.

14. Symposium on "Safer Cancer Chemotherapy". Royal College of
        Surgeons of England. 1981.
        Eds: Price, L.A., Hill, Bridget T. & Ghilchik, M.W.
        Publ. Bailliere Tindall.

15. Tate, H.C., Rawlinson, J.B. & Freedman, L.S. 1979.
        Lancet, 2, 623.

16. Hill, Bridget T. 1980.
        Principles of Tumour Growth.
        In: Scientific Foundations of Respiratory Medicine.
        Eds: Scadding, J. G. & Cumming, G. p.593.

17. Price, L. A. 1980
        Principles of Tumour Growth.
        In: Scientific Foundations of Respiratory Medicine.
        Eds. Scadding, J. G. & Cumming, G. p.603.

18. Carter, S. K. & Soper, W. T. 1974.
        Cancer Treatment Reviews, 1, p.1.

19. Hill, B. T. & Price, L. A. 1977.
        Concepts and Prospects in Adjuvant Chemotherapy.
        In: "Secondary Spread in Breast Cancer";
        Vol. 3. New Aspects of Breast Cancer, p. 193.
        Ed: Stoll Heinemann Medical, London.

20. Price, L. A. & Hill, Bridget, T. 1974.
        Classification of anti-tumour drugs.
        Lancet, ii, 172.

DISCUSSION

SPEAKER: PRICE   **CHAIRMAN: CUMMING**

FABBRI: I'd like to know your opinion on surgery in small
                cell lung cancer in the absence of clinically
                detectable metastatic lesions.

PRICE: The present consensus of opinion in the United
                States is that the cure rate for surgery alone in
                small cell lung cancer is less than 5% and even
                then largely confined to physiologically fit
                patients with peripheral lung lesions. Therefore
                one cannot exclude surgery totally in this disease
                but it has a very limited role and certainly
                should not take precedence over drug treatment or
                even radiation therapy.

CHAIRMAN: I'd like to take the opportunity to present a
                preliminary report on a controlled trial of these
                methods of therapy in carcinoma of the bronchus.
                The trial is unfortunately small in size but there
                is a five year follow up and I can report that
                patients who presented with carcinoma of the
                bronchus were randomly allocated to two groups.
                These were those who had surgery alone, the
                surgery consisting either of lobectomy or, in a
                small number of cases pneumonectomy and those who
                received in addition to surgery, adjuvant
                chemotherapy of the type we have just heard about
                at proper intervals. One complication of random
                allocation is that one cannot allocate randomly on
                the basis of cell type so that the cell type
                structure of the two groups was not exactly the
                same. Large cell tumours dominated in both groups
                but in the control group there were fewer large
                cell tumours than in the experimental group. The
                results however were somewhat surprising: in a
                group of 10 patients who received adjuvant
                chemotherapy, one patient died from lung fibrosis,
                brought about by the bleomycin and one person was
                unable to complete the full course of adjuvant
                chemotherapy because of intolerance. This patient
                died after about 18 months from disseminated
                disease. However 8 out of the remaining 10 are
                currently alive and well and living normal lives

five years after therapy. If one compares this 80% survival with the 10% which is normally seen it is obviously a significant improvement. However in the control group where one might have expected a similar value of 10% we had in fact a value of 40%. This is very strange and perhaps we are looking at different populations for example a rural population rather than an urban one or possibly the method of surgery was different in some way, so two groups were observed and the survival rate in the control group was surprisingly good but that in the adjuvant chemotherapy group was magnificent i.e. 80% survival at five years.

SPINA: I'd like to know your opinion on the possible reduction of the lung tumour volume before starting chemotherapy in non-operable patients. Our experience has not been very successful in this respect and I wonder what your opinion is about that.

PRICE: I agree with you that in the studies of which I am aware, where attempts to reduce the tumour volume by an initial attack using radiation have not in general been successful. However there are two studies which might provide a little encouragement. One is the adjuvant use in non-small cell carcinoma or combinations including mitomycin C, vinblastine and cis-platinum which I am told is showing encouraging results in an adjuvant study although this has not been published yet. Another study by Anderson in Wales where chemotherapy was given immediately after radiation using the 24 hour approach that I have described showed a significant increase in survival in the chemotherapy group compared with the radiation only group. This study is more important than it seems because the chemotherapy was not by modern standards very intensive and it does suggest that we might be able to do better in the future if we (a) give chemotherapy immediately after (or even before) radiation and (b) if we give modern intensive chemotherapy as outlined in my lecture.

MARINI: Two questions; you mentioned VP16-213 amongst the most active phase-specific drugs in small cell lung cancer. Have you got any experience with the association between cis-platin and VP16-213? The second question concerns the use of metoclopramide as an anti-emetic drug. How effective is it with the high doses you mentioned and which are the main side-effects.

PRICE: VP16-213 (Vepesid) is the single most effective drug in small cell lung cancer and can be given much more safely by infusing it over 24 hours instead of the suggested administration over 3 or 5 days. My scientific colleague Dr. Bridget Hill at the Imperial Cancer Research Fund is investigating the experimental basis for interactions between cis-platin and VP16-213. I am not at liberty to tell you what these results are but they will be available shortly. The present evidence clinically is that cis-platin should probably be used only in squamous cell carcinoma although I personally would like to see more evidence before using this drug in any type of non-small cell lung cancer. The high dose metoclopramid was a method of attempting to overcome severe nausea and vomiting associated with cis-platin therapy. Even when given over 24 hours as Dr. Hill and I advocate it still produces these distressing side-effects. Recently however it has been shown that 120 mg of metoclopramide in 50 ml of normal saline infused over 15 minutes starting 1.1/2 hours before the cis-platin infusion and repeated every 2 hours significantly reduces nausea and vomiting. There are extra-pyramidal side-effects in some cases. Recent evidence reported at the St. Louis meeting of the American Society of Clinical Oncology this year suggest that equally good anti-emetic effects can be produced by using half these doses, i.e. 75 mg instead of 120.

BONSIGNORE: What is your view of the role of laser therapy in the treatment of lung cancer?

PRICE: I'm afraid I do not know much about this method of treatment but no doubt it has some value in certain cases. I still think however that the basic principle holds that it is still a local kind of treatment for what is a systemic disease. My personal view is that we need more information about this.

CHAIRMAN: If I could make a brief comment on the role of
                laser therapy, I think this is largely where the
                tumour mass is so occluding the bronchus that it
                threatens life from strangulation and then it is
                relatively easy to fry the tumour with lasers and
                they shrink markedly but I think Price is telling
                us this does not attack the systemic dissemination
                which by definition is present at that time. This
                may of course be one of the two ways of tackling
                the inoperable carcinoma that Professor Spina
                talked about since carcinomas are generally
                inoperable when they involve the carina. If the
                carina can be attacked with laser therapy and
                adjuvant therapy given at the same time that may
                be a way of combining these two types of therapy.

PRICE: May I just say Mr. Chairman that I do not wish in
                any way to discredit local therapy. What I wish
                to see is the best kind of local therapy
                integrated with the best kind of systemic therapy.

ROSSI: You distinguish in your talk between extensive and
                limited disease and you say that's important for
                prognosis. There has been a study last year on
                the American Review of Respiratory Disease by
                Giampina and he says that its much more important
                in terms of prognosis that there is a good
                performance status and whether or not there are
                metastases in the liver and in the brain. Would
                you comment on that?

PRICE: When I said that limited disease was important I
                was speaking about small cell lung cancer where in
                "limited" disease (i.e. disease confined to one
                hemi-thorax with or without the draining lymph
                nodes) there is a significant chance of achieving
                an 18 to24 month disease free remission compared
                with extensive disease (i.e. where there are overt
                metastases) where there is nothing like such a
                good chance. I agree with you that ambulatory
                status is important and it is also a good
                prognostic factor in other tumours as well. I am
                sure that the sites of metastases obviously are
                important and make a difference to the prognosis
                although I don't have the data for each individual
                site.

ROSSI: Giampina is saying that the most important sites
                are liver and the brain. He spoke about

prophylactic irradiation of the brain because its a place where it is difficult to achieve good drug concentrations but he didn't mention anything about the prophylactic irradation of the liver. Do you think that's important or do you think that chemotherapy can do a good job in the liver?

PRICE: I am still talking about small cell lung cancer and I think prophylactic irradiation of the brain should be delayed until you know if the patient is going to achieve a complete response following two courses of chemotherapy (i.e. within 6 to 8 weeks). I think this position is ethically defensible because the brain metastases do not usually become clinically obvious until about 12 to 14 weeks. By that time you know if the extra-cranial tumour is responding and if it is I think prophylactic cranial irradiation should be carried out. However, if the extra-cranial tumour is not responding I think its rather pointless to irradiate the brain since this is a very unpleasant procedure for a patient whose prognosis is not very good. I think that in small cell carcinoma which has a 70% response rate to chemotherapy one expects liver secondaries to respond and I would oppose routine prophylactic liver irradiation which also makes people feel worse.

QUESTION: The connection between kinetics and therapy is doubted by many investigators, would you comment on that.

PRICE: It is obvious that I have not made clear that the kinetics I was talking about refer only to the stem cell compartment. I think that the old fashioned kinetics about heterogeneous measurements on heterogeneous cell populations such as doubling time, labelling index etc. are totally clinically irrelevant. I was speaking only about the kinetics of the sub-fraction of stem cells i.e. homogeneous cell kinetics. At the risk of becoming unpopular I cannot understand why many writers on this subject seem unable to make this extremely simple distinction. The whole point about the application of the Bruce stem cell kinetic model is that over a period of 24 hours fewer bone marrow stem cells are cycling and more tumour stem cells are cycling and therefore therapy confined to this period will always selectively damage more tumour stem cells and fewer normal bone marrow stem cells. Bruce has also shown that if you extend the time of exposure of the drugs to several days this selectivity is completely abolished and this is also what is observed clinically. I also think that it is great error to extrapolate data from leukaemia to solid tumours. The theory that I mentioned is based on the assumption that the marrow is not involved and has normal marrow stem cells kinetics whereas in leukaemia this may not be the case. I am not an authority on leukaemia, but in the few cases I have treated the use of the 24 hour principle significantly reduces damage to the normal marrow.

CORRIN: May I ask a rather naive question. Could you tell us more about how these drugs kill cancer cells. Your whole emphasis seems to be on kinetics and therefore the anti-mitotic activity. Do they kill tumour cells in any other way? If not perhaps tumour cells and bone marrow cells have common cells dividing and there should perhaps be an optimum chemotherapy for all forms of cancer. But we all know very well that the different forms of cancer respond to different forms of chemotherapy, they are all dividing. Is there another way that these drugs are killing the cancer cells?

PRICE: I will answer that question as briefly as I can because although you put it very tactfully, it is extremely complex. First, anti-tumour drugs have different effects on tumour cells and this is often dose dependent. For example they can hold cells up in a specific phase of the cycle such as M or G1 or they can kill cells at the same phase of the cycle if you give a bigger dose, for example vincristine and vinblastine in small doses inhibit mitosis but kill cells in bigger doses in S phase. Your statement again about all cancer cells dividing avoids the point that I was only talking about stem cells and the fact that they are probably dividing more rapidly or to be more exact more of them are dividing over a 24 hour period than are the bone marrow stem cells most of which are in G0 during this time. It is not true that all cancer cells divide rapidly. As a matter of fact if you take these old fashioned

MULTI-DISCIPLINARY TREATMENT FOR LUNG CANCER   153

heterogeneous measurements such as doubling time, some of the differentiated cells as they get towards the end of their genetically determined number of divisions are hardly dividing at all. Finally I am not saying that you cannot achieve good results by giving drugs over five days, I am only saying that it is better and safer to give them over 24 hours and that where we have done this we have got increasing safety without loss of therapeutic effect. Our theory is a comprehensive one i.e. we say you should give drugs over 24 to 36 hours for all tumours, but of course you must use different drugs because there are different sensitivities in different tumours and some drugs are effective as you say in some tumours and not effective in others.   It would be futile for example to treat lung cancer with busulphan as the MRC did a few years ago because it doesn't work in lung cancer and that trial' proved what could have been predicted without the trial being carried out at all.   In fact the patients receiving busulphan had a worse quality of life than those who had no chemotherapy at all.   This merely illustrates yet again the obvious point that if you use ineffective chemotherapy you will achieve ineffective results.

TREATMENT OF PULMONARY TUBERCULOSIS

Giuseppe Spina

University of Palermo, Sicily

Chemotherapy in pulmonary tuberculosis, has undergone with the passing of years great changes because of the discovery of new antibiotics, and also because of a more precise identification of their effectiveness in terms of ability and speed in sputum conversion and to the ability to induce recovery and avoid relapses. Therefore in the anti-tubercular therapy we must consider the following points:

- Effectiveness of drugs

- Route of administration

- Recovery time

- Clinical and bacteriological relapses

- Toxicity

- Cost/benefit ratio

Effectiveness of antitubercular drugs

The evaluation of the effectiveness of a drug is based on the following parameters:

- Minimum inhibiting concentration (M.I.C.)

- Proportion and nature of resistant mutants

- Serum concentration in relation with M.I.C.

–   Effect on experimental tuberculosis.

The minimum inhibiting concentration (M.I.C.) is the lowest drug concentration able to inhibit the formation of M.Tub for which the lower the M.I.C., the more effective is the drug: i.e. for isomazid (H) it is of mcg 0.04/0.07, for rifampin (R) it is 0.15/0.30, for pyrazinamide (Z) it is 10.

The proportion of spontaneously resistant mutants is also important because this increases the risk that during therapy there may be formation of resistant M.Tub. This proportion is calculated from the number of resistant colonies, among the sensitive ones, that is able to grow at the M.I.C. of that particular antibiotic. For R this value is of 1×10^{-8}, for H and Streptomycin (S) of 1×10^{-6}.

The nature of resistant mutants concerns the concept of cross-resistance: it exists between the aminoglycoside antibiotics but in an incomplete form. That existing between streptomycin–kanamycin acts in only one way, in that the strains resistant to kanamycin are also resistant to streptomycin and not the other way around. Another feature that must be considered is that of the reduced virulence of the resistant strains already shown for H by Middlebrook in 1953 and proportional to the intensity of the resistance. This phenomenon has also been demonstrated for resistant R strains.

The serum concentration that can be reached by the antibiotic is important because the higher it is the greater is its effectiveness particularly when a high serum concentration (s.c.) corresponds to a low M.I.C. For example for 450·mg of H we reach s.c. of 2 mg at the third hour, a value 100 times higher than the M.I.C., and a 600 mg dose of R gives at the third hour an s.c. of 9 mg which is 75 times higher; for X mg streptomycin it is 30 times higher;   while for X mg pyrazinamide it is only 5 – 10 times higher.

Effect of antibiotics on experimental tuberculosis: is evaluated from the speed and stability of the conversion of tuberculosis in the mouse, evaluated with the culture of the lung and of the spleen. For example the association HR converts by 100% in 5 months and by 70% after a follow up of 4 months, while HS by 50% after 9 months and by 0% after a follow up of 6 months respectively. We must bear in mind that these data, although highly significant, are far from what happens in man who is capable of offering resistance to the infection and contributes to the speed of the recovery and to the elimination of the remaining bacteria and thus to the low number of relapses.

On the basis of the above concepts Grosset has given a classification of drugs active on M. tuberculosis (Table 1).

Therapy of tuberculosis

In planning a therapy for tuberculosis we must bear in mind the following aims of chemotherapy:

- Fast conversion and recovery

- Prevention of resistance

- Prevention of relapses

- Avoidance of toxicity

A fast conversion and recovery depend on the use of the major drugs according to the effectiveness just mentioned. The speed of the recovery is closely related to the prevention of resistance: this aim may be attained with the combination of antibiotics proved to be effective. The combination of antibiotics is also effective against possible primary resistances such as the one that may be found in patients who have never been treated with specific antibiotics. Generally the resistance is towards one drug only and more frequently in the following order S, H: in 1974 in our patients it was 6.85% and 3.22% respectively, while for R and E the phenomenon is rare (in 1974 0.4½ and 0.8% respectively.

In Table 2 we report values of primary resistance found in various countries.

This investigation was deemed necessary in the past at the beginning of therapy: but considering the use of two or three drugs today it is no longer necessary. A controlled study in Hong Kong (Fox) showed independently that the therapeutic result was the same.

The avoidance of toxic phenomena is very important because a lower toxicity enables a prolonged use of an antibiotic without damage to the patient. The toxicity for the drugs used more widely is reported in Table 3.

Although each antibiotic must be used in its therapeutic dose the combination of drugs tends to reduce toxicity in that it causes a faster recovery, reducing the period of therapy. Nevertheless an increased toxicity has been observed from the

TABLE 1

CLASSIFICATION OF DRUGS ACTIVE ON M. TUBERCULOSIS

High activity	Isoniazid Rifampin
Medium activity	Streptomycin Kanamycin Pyrazinamide Ethambutol Ethionamide Cycloserine
Low activity	P-aminosalycilic acid Thiacetazone

J. Grosset - Rev. Fr. Mal. Resp.
1976, 4, 157

TABLE 2

OCCURRENCE OF PRIMARY RESISTANCE TO M. TUBERCULOSIS

COUNTRY	YEAR	%
U.K.	1955 1963 1970	4,5 4,1 5,4
U.S.A.	1962 1967 1973	9,7 8,7 10,0
France	1961 1967 1971/74	9,3 9,7 8,8
Italy (Rome)	1971 1972 1974	10,0 9,8 10,0

TABLE 3

TOXICITY OF DRUGS ACTIVE ON M. TUBERCULOSIS

DRUG	DOSE mg	MAIN TOXICITY	SYMPTOMS
S. Sulfale S. Dihydro	1000	Vestibular and auditory branch of 8th-kidney	Dizziness - Vertigo Hearing loss transient cylindruria
H	6-8/kg	Liver, CNS, PNS	Dispepsia - Liver enzyme disturbances - Neuritis - Emotional disorders
R	12-15/kg 600/900	Liver - Stomach Platelets	Jaundice - Liver enzyme disturbances - Nausea - Vomiting - Diarrhoea - Thrombocytopenia - Haemorragic purpura (rare)
E	25/kg 1500-2000	Optic N.	Axial retrobulbar neuritis
Z	1500-2000	Liver	Jaundice - Liver enzyme disturbances - Hyperuricoemia

association of INH + RMP or for RMP in two or three week therapies. Also Pz associated with to the two previous drugs increases therein toxicity. Hepatic toxicity being the most frequent form one should give particular attention to patients who are at risk from liver disease.

The prevention of relapses is possible by the use of antibiotics with a stronger bactericide action. A long series of clinical and experimental investigations have pointed out the importance of the action of H + R with the eventual help of a minor drug among which Z is very important.

The therapeutic regimes for the fulfillment of the aims described above are many and can be summarised as follows:

-    long term continuous therapy (12-18-24 months)

-    intermittent therapy (without time limits)

-    short course therapy (12 months - suggested 6 - 9 months

Long term continuous therapy

Several studies carried out with three drugs have demonstrated the effectiveness of long term continuous therapy in the treatment of pulmonary tuberculosis. In the past they were SM, INH, PAs administered for at least 18 months. Nevertheless long term therapy caused toxic phenomena 13.8% for PAS after 4 months and 39.8% after 8 months (Gyselen et al, 1962). Also SM often caused cochleovestibular toxic phenomena appearing mainly after 50 - 60 gr. In the best situations relapses were 9% - 19% if treatment was suspended after about one year and 4% if therapy was extended for 2 - 3 years (M.R.C. 1962). As regards the effectiveness of PAS the chemotherapy centre of Madras, having compared SM + INH treatment with and without PAS, observed a bacteriologic conversion of 85% and 87% respectively.

In the cited MRC investigation, comparing a treatment based on INH and PAS with and without SM 1 g daily during the first six weeks, a positive result was found in 83 and 75% respectively: this difference is only moderately significant.

Ethambutol, because of its lower toxicity, has gradually replaced PAS with an improvement in the results.

A real turning point was reached in 1966 with the introduction of ritampicin (R) and the need for such long therapies was questioned by several experts. We have also spoken of intermittent therapy and of short course therapy in relation to the speed of conversion obtainable with schemes including R.

Table 4 shows the results obtained by Nitti et al in 1972.

Intermittent therapy

Dickinson and Mitchinson demonstrated that M.Tub. after exposure to antitubercular antibiotics shows a prolonged lag period: no change for 6 hours after contact of M.Tub. with H and E, while for 24 hours the lag period is of 6 - 10 days for H and 3 - 5 for E; for R, 0.2mcg/ml, the lag period after 6 hours is 2 - 3 days.

The experimental research of Grunbach et al. 1969 of the Pasteur Group of Paris demonstrated the findings reported in Table 5.

It can be seen that the most effective therapy is the one that, either in daily or intermittent administration, is associated with RH and that the twice a week intermittent treatment for a period of 5 months is clearly more effective than the twice a week intermittent therapy for four months. In the follow up it can be seen that the addition of S brings about only a modest improvement while the total duration of 5 months is less favourable than that of six months.

The Italian investigators (Lucchesi - Nitti) have also produced positive contributions to intermittent therapy on the experimental level as well as McCune and Tompsett, Bartmann.

Clinical experiments were started at the Madras Center in 1959 and after that by many authors (Daddi, Lucchesi, Nitti, Mesti Decroix, Kreis, Mallorquin, Durba, Verlist, Ziersky, Larbaoui, Chaulet, Farga, Gomi, Giobbi).

Table 6 reports the bacteriological results obtained by various authors.

Pretreatment seems to be useful for the conversion of sputum for M. Tub., but its duration is less significant. The results of the one-week treatment are not good. The clinical results following various treatments including mainly SH seem to be less satisfactory; at the end of 12 months the cavity was closed inbetween 24% - 38% of patients and a result deemed "excellent", in 94 - 92%.

TABLE 4

MEAN DURATIONS OF THERAPY INDUCING
CONVERSION OF SPUTUM

THERAPEUTIC SCHEDULES	TIME
RIFAMPIN + ISONIAZID (1ST EXPERIENCE)	44.4 ± 3.2
RIFAMPIN + ISONIAZID (2ND EXPERIENCE)	44.1 ± 2.8
RIFAMPIN + ISONIAZID + ETHAMBUTOL (2ND EXP.)	45.6 ± 3.1
RIFAMPIN + ISONIAZID + STREPTOMYCIN (2ND EXP.)	43.4 ± 3.2
ISONIAZID + STREPTOMYCIN + ETHAMBUTOL (REFERENCE TREATMENT 2ND EXPERIENCE)	69.1 ± 4.8

TABLE 5

NEGATIVE COLTURES OBSERVED IN MICE AFTER
DAILY AND DAILY + INTERMITTENT THERAPY

SCHEME	REGIME	NEGATIVE AT 6 MOS. %		NEGATIVE 4 MOS. AFTER COMPLETION OF THERAPY %	
		LUNG	PLEEN	LUNG	SPLEEN
DAILY 6 MOS.	RH	100	100	-	-
	RE	75	87	-	-
	HE	9	0	-	-
DAILY 5 MOS.	RHS	100	100	92	77
	RH	100	100	80	70
DAILY 1 MOS. THEN TWICE/WEEK FOR 5 MONTHS	RH	100	78	-	-
	RE	37	37	-	-
	HE	12	0	-	-
DAILY 1 MOS. THEN TWICE/WEEK FOR 4 MONTHS	RSH	70	90	64	64
	RH	70	60	66	50

(GRUMBACH ET AL., TUBERCLE, 1969, 50, 280)

TABLE 6

BACTERIOLOGICAL RESULTS AFTER 12 MOS.
PRIMARY INTERMITTENT REGIME

INVESTIGATION	DAILY PHASE 1-3 MOS.	INTERMITTENT PHASE	PATIENTS	NEGATIVE AT 12 MOS. %
Uict 1970, Polansky 1970	IES	SH_2	639	94,6
Larbaoui, 1970, Lahlou 1970	IES	SH_2	167	90 (8 mos.)
Who 1971, Devi 1972				
Madras 1970	NO	SH_2	117	91
Menon 1970	IES	SH_1	181	85

S = Streptomycin   1 = once/week
H = Isoniazid   2 = twice/week

Intermittent therapy with R, carried out in socially and economically developed countries, has given very good results as can be seen in Table 7.

The clinical results are shown in Table 8.

Tolerance is generally good with jaundice in 2.41% of the cases on average. In intermittent therapy with R there have been cases of fever and thrombocytopenia owing to the appearance of anti R antibodies. In a survey by Binda on 11,631 cases, 3 cases of purpura and 16 haemorrhages were observed. Leucopenia was found in 13 - 14% of the cases on average.

In reports by Basi et al the incidence of hypersensitivity reactions was 3.6% if the intermittent therapy was preceded by daily dose and 9.8% if the dose of R was 900 mg. At the UITC Congress (Bruxelles, 1978) for an R dose of 12 mg/K fever and temperature were observed in 1.1% of the cases and anti R antibodies in 0.9%.

In short, considering the data in literature, the following conclusions can be drawn:

1. an initial continuous regimen of 1 - 3 months is desirable.

2. The drug dosage, during intermittent regimens, should be higher than usual and at single doses.

3. A twice weekly regimen is preferable.

4. Treatment must be continued until healing has taken place..

5. Social and economic factors permitting rifampin should be present in every scheme of treatment.

Short course therapy

According to the definition accepted by the UITC treatment committee in Istanbul 1977 and in Bruxelles 1978, short course are "all those continuous therapies for a period under 12 months" with an ideal duration between 6 and 9 months.

This kind of therapy in the past few years has met with increasing favour both in the developing countries and in developed our countries. It is based on the experimental studies of the group of the Paris Institute Pasteur and Mitchinson in U.K. and on many controlled therapeutic assays carried out in several countries (British Thoracic Association,

TABLE 7

BACTERIOLOGICAL RESULTS OF INTERMITTENT REGIMENS

INVESTIGATION	DAILY (CONTROLS)	INTERMITTENT				NEGATIVE AT 6 MOS %
		DAILY PHASE	INTERMITTENT PHASE			
			RH_1	RH_2	RH_3	
Doyle 1969 Mallorquin 1970 Silveira 1970	RH (116)	NO (145)	(76)	(69)	–	100
						100
Poole 1971 Decroix 1971 Verbist 1971 Dubra 1972		1-3 RH_7 (205)	(18)	(169)	18	100

Daily phase   R mg 600 – H mg 10/Kg
Intermittent phase   R mg 900 – H mg 15/Kg

TABLE 8

CLINICAL RESULTS AT 12 MOS. OF PRIMARY
INTERMITTENT REGIMEN INCLUDING RIFAMPIN

INVESTIGATORS	REGIMEN		No. CASES	OUTCOME			
	DAILY PHASE	INTERM. PHASE		DIS. CAVIT.	GOOD	MODER.	POOR
Decroix-Kreis 1971	RH_7		39	86	?	?	5,6
	NO	RH_2	41	87	?	?	4
Dubra 1972	SH	RH_2	69	73 (6 mos.)	?	?	?
	RH_7		67	73 (6 mos.)	?	?	?
Verbist 1970	NO	RH_1	48		62	35	2,1

Spina, Riv. Sic. Tuberc., 1973, 27, 472

East African BMRC, Hong Kong BMRC, Tuberculosis Research Centre Madras, Brouet and Roussel in France). While in standard therapy one must look for the combination which inhibits the onset of resistance, drugs being administered for the time necessary to eliminate the bacterial population and attain recovery. In short course therapy one must use drugs with a fast sterilising action and therefore drugs which indirectly hinder bacterial resistances. This, according to Fox, is judged from the conversion obtained in the first two months of therapy and from the rate of clinico-bacteriological relapses.

Bacteriological research (Mitchinson, 1980) has shown the presence of four classes of bacilli according to the scheme in Table 9.

The bacilli of the first group which grow rapidly (cavity wall) are destroyed by H, R, S, but particularly by H. Among these bacilli there is a population which has a slower and sometimes intermittent growth and they are destroyed mainly by R but also by H though to a lesser extent; a population with a very slow growth in acid pH and is found inside macrophages and cavities and is killed very well by Z which acts in an acid environment. There is finally a population of sleeping bacilli which is attacked by no antibiotic, but with the passing of time may cause relapses.

Grosset (1978) summarises the experimental data as follows:

- Z and R are effective on experimental infection in mice.

- H + Z and H + R show the highest effectiveness.

- Adding S or E causes low - if any - increase in the effect of associations.

The results of a short course therapy are reported in Table 10.

In short course therapy one tends today to reduce the initial time of therapy with more than one drug (3 or 4) to 2-3-4 months and to continue with therapy using two drugs or with one drug only. The results are very satisfactory as can be seen in Table 11.

In a recent bacteriologic and radiologic study reported by Chen and Girling on Hong Kong patients with negative sputum and positive culture and both negative, it has been pointed out that a 2 or 3 months treatment with up to 4 drugs was unable to prevent clinico-bacteriologic relapses.

TABLE 9

BACTERIAL POPULATION AT THE LEVEL OF THE
LESION AND ACTION OF ANTIBIOTICS

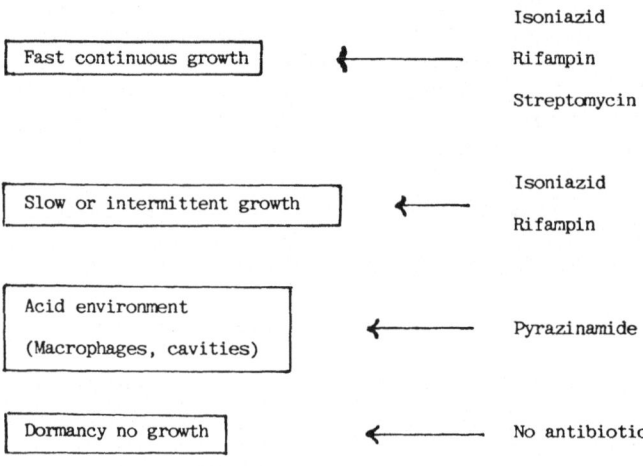

| Isoniazid |
| Fast continuous growth ← Rifampin |
| Streptomycin |

Slow or intermittent growth ← Isoniazid / Rifampin

Acid environment (Macrophages, cavities) ← Pyrazinamide

Dormancy no growth ← No antibiotic

Mitchison (modified) 1980, J.R. Coll, Phys, 14, 91

TABLE 10

SHORT COURSE CHEMOTHERAPY IN FRANCE AND U.K.

	TREATMENT	DURATION OF THERAPY IN MOS.	PATIENTS	RELAPSES	DURATION OF CONTROL AFTER THERAPY IN MOS.
France (Brouey et al. Roussel 1975)	SHR	6	66	3	30
	HR	9	74	0	27
	EHR	12	69	0	24
U.K. (British Thoracic and Tuberculosis Association) (Angel et al. 1975)	SHR	9	151	0	18
	HR	12	177	1	15
	EHR	18	155	0	9

S = Streptomycin   -   H = Isoniazid   -   R = Rifampin

TABLE 11

THERAPEUTIC REGIMEN SHOWN AS EFFECTIVE
ON 7-9 MOS. COURSES ON SPUTUM-POSITIVE PATIENTS

INVESTIGATION	INITIAL THERAPY	FOLLOWING THERAPY	DURATION IN MONTHS	PATIENTS	RELAPSES	CONTROL IN MONTHS
2BTA, Brouet e Roussel, Pretet (1977 - 81)	2-3 SHR 2-3 EHR	HR	9	298	1	9-45
Lees (1981)	2-4 RHE	HE	9	82	0	12
East African BMRC (1980)	2 SHRZ	HT	8	81	0	22
Aluoch (1981)	2 SHRZ	H	8	102	1	6
Tuberculosis Research Centre Madras (1981)	2 SHRZ	SHZ_2^*	7	132	2	41

(W. Fox, Bull. UITC, 1981, 56, 147-169)

In conclusion, considering the data available, we can say that a 6 months regimen seems to be the more advisable for patients with a positive culture. Patients with positive culture only may need a 5 or even 6 months treatment: for patients with negative culture 4 - 5 months are probably enough. For patients with positive sputum a further treatment may prove useful, 6 months after recovery as a prophylaxis for relapses.

We think that in socially and economically advanced countries we can today perform a <u>personalised therapy</u> taking into account the suggestions of the reported investigations. According to the severity of the illness, its extent (uni- or bi-lateral, with single or multiple cavitations, with a large inflammatory component) the therapy may be carried out for 4-5-6 months or more on a daily basis, followed, after recovery, by a 2 - 3 months intermittent therapy twice a week with two drugs, RH or HE for example. In less severe cases the same intermittent therapy can be followed with H only.

The large use of rifampin both in intermittent and short course treatments must not overlook the possibility of toxic phenomena particularly for hepatic function. There is a difference in the number of jaundice cases between the Italian case reports and the foreign ones as can be seen in Table 13 is concerning polytherapy with R and H.

The association R + H has turned out to be the more toxic, but toxic phenomena disappear rapidly after therapy. However a good selection of the patients, based on their life habits (drinking, chronic liver diseases, etc) reduces toxicity noticeably. On the whole, in subjects with normal hepatic functions, asymptomatic hepatic alterations can be observed (serum transaminases, bilirubin, alcaline phosphatase levels) which disappear during the course of therapy or temporary suspensions: the same can be said for the cases of jaundice where at least a temporary interruption is necessary. The short course treatment based on HRZ gives acceptable toxicity values (Fox, 1980) and does not usually condition either the therapy or its acceptance by the patient.

The incidence of hepatitis during various treatments is shown in Table 14.

Corticosteroid therapy associated with antitubercular chemotherapy has been found useful by many.

The association causes a rapid improvement of general

TABLE 12

RELAPSE RATES IN TWO INVESTIGATIONS IN HONG KONG

EXPECTORATION	DURATION OF THERAPY IN MOS.	DURATION OF FOLLOW-UP IN MOS.	PATIENTS ASSESSED	RELAPSES				
				BACTERIOLOGICAL		ALL CATEGORIES		
				No.	%	No.	%	CONFIDENCE LIMITS 95%
NEGATIVE SPUTUM AND POSITIVE CULTURE	2SHRZ	46	69	13	19	19	28	17-40
	3SHRZ	45	68	6	9	8	12	5-22
	4SHRZ	8+	78	3	4	3	4	0.8-11
	4SHRZ3	8+	63	0	0	0	0	0-6
	6SHRZ3	6+	81	0	0	2	2	0.3-9
NEGATIVE SPUTUM AND CULTURE	2SHRZ	46	154	9	6	16	10	6-16
	3SHRZ	45	154	3	2	8	5	2-10
		9+	159	0	0	3	2	0.4-5
	3SHRZ3	9+	181	2	1	6	3	1-7
	4SHRZ3	8+	146	0	0	1	1	0.02-4
	SELECTIVE2	48	171	71	42	96	56	49-64

CHEN 1981 - GIRLING 1981

TABLE 13

SERIES	No. CASES	JAUNDICE	%
ITALY	1346	19	1,41
ABROAD	1421	58	4,08

TABLE 14

INCIDENCE OF HEPATITIS DURING STANDARD
DURATION ISONIAZID TREATMENT ASSOCIATED
WITH ONE OR MORE DRUGS (B.M.R.C.)

COUNTRY	ASSOCIATION	No. CASES	HEPATITIS	
			No.	%
U.K. (2 INVEST.)	SHP	444	5 (5)	1
	SHE	116	1 (0)	1
	SHR	112	5 (1)	4
INDIA (5 INVEST.)	SHP	519	9 (9)	2
	SH	78	0	0
	HT	75	2 (2)	3
EAST AFRICA (5 INVEST.)	SHT	1665	10 (10)	1
	HP	125	2 (0)	2
HONG-KONG	HP	100	1 (1)	1
	HT	150	8 (8)	5
SINGAPORE	SHT	225	8 (6)	4
	SH	131	2 (1)	2
INTERNATIONAL (2 INVEST.)	SHT	3563	15 (2)	0,4
	SH	2394	5 (1)	0,2

GIRLING D.J., BULL. UITC, 1980,55

TABLE 15

CLOSURE OF CAVITIES IN 180 PATIENTS

THERAPY	AFTER 6 MOS.		AFTER 12 MOS.	
	PLACEBO	STEROIDS	PLACEBO	STEROIDS
MEDICAL	22%	25%	40%	48% N.S.
SURGICAL	50%	83%	30% S.	50% S.

JOHNSON ET AL., AM.RESP.DIS.,92,3,376,1965

TABLE 16

POSSIBLE CAUSES OF FAILURE IN
THE TREATMENT OF PULMONARY TUBERCULOSIS

- IRREGULAR UPTAKE
- POOR ABSORPTION
- LACK OF PENETRATION AT THE LEVEL OF LESION
- ACQUIRED RESISTANCE
- TOO SHORT DURATION OF TREATMENT
- POOR DEFENSE MECHANISMS
- SEVERITY OF LESIONS
- INTOLERANCE TO DRUGS

CROFTON J., BULL. UITC, 55, 1980

conditions, a drop in temperature, a faster regression of the inflammatory component in the lesions according to Johnson et al., but a non-significant closure of cavities. These authors, in a controlled study with the use of R, have observed the findings reported in the following Table 15.

In another controlled study (Horne) no significant difference was observed between subjects treated and those who were not treated with corticosteroids. But this study antedated the use of R.

The variable doses are between 30 - 40 mg prednisolone: these doses can be reduced after an improvement has been attained. One must note that if corticosteroids are suspended too soon there may be a rebound phenomenon.

Finally in the chemotherapy of pulmonary tuberculosis there may be some failures the causes of which are described in the Table 16 and which Crofton (1980) has recently called our attention.

REFERENCES

Binda, G., Domenichini, E., Gottardi, A., Orlandi, B., Ortelli, E., Pacini, B., Fowst, G. 1971.
Rifampicin, a general review. Arzneim - Forsch. 21, 1907-1977.

Blasi, A., Donatelli, L., Zanussi, C. 1978.
Rifampicina 315-330, Ed. Minerva Medica.

Blasi, A., Nitti, V., Delle Veneri, F., Virgilio, R., Marsico, A. S. 1973.
A controlled clinical evaluation of 3 intermittent chemotherapy regimens employing low dosage schedules of rifampicin in the treatment of newly diagnosed cases of pulmonary tuberculosis.
XXII Int. Tuberc. Conf. Tokyo, 20-29.

Chen. Girling Bull. UITC, 1981, 56, 157. (Reported by Fox).

Crofton, J. Bull UITC v.55, September-December, 1980.

Grumbach, F., Canetti, G., Le Lirzin, M. 1969.
    Rifampicin in daily and intermittent treatment of
    experimental neurine tuberculosis, emphasis on late
    results.
    Tubercle Lond., 50 280.

Gyselen, A., Casemans, J., Vandenbergh, E., Laquet, L. M.,
    Verbist, L. 1971.
    Effets secondaires de la chemiotherapie anti
    tuberculeuse: douze annees d'experience sanatoriale.
    XXI Int. Tuberc. Coference, Moscow. Excepta medica 74.

Horne, N. W. 1960.
    Prednisolone in treatment of pulmonary tuberculosis: a
    controlled trial. Final report to the Research Committee
    of the Tuberculosis Society of Scotland.
    Brit. Med. J., 2, 5215, 1751-1756.

Johnson, J. R., Taylor, B. C., Morrissey, J. F., Jenne, J. W.,
    Macdonald, F. M. 1965.
    Corticosteroides in pulmonary tuberculosis.
    Am. Rev. Resp. Dis., 92, 3, 376-391.

Lucchesi, M. 1971.
    La terapia intermittente antitubercolare basi
    sperimentali.
    Annali Medici de Sondalo, 19, 125.

Mitchison, D. A., Dickinson, J. M. 1978.
    Mecanisme de la bactericidie dans la chimioterapie de
    courte duree.
    Bull. UITC, v. 53, 4, 270-275.

Report Medical Research Council by their Tuberculosis
    chemotherapy trials Committee. 1962.
    Long term chemotherapy in the treatment of chronic
    pulmonary tuberculosis with ????.
    Tubercle, 43, 201

Spina, G. 1970.
    Intolleranza nella terapia della tubercolosi polmonare
    trattata con Rifampicina.
    Simposio Nazionale sulla rifampicina.
    Roma Settembre, (Simposi Lepetit).

Spina, G. 1978.
    Rifampicina, 229-230.
    Ed. Minerva Medica.

DISCUSSION

SPEAKER: SPINA   **CHAIRMAN: BONSIGNORE**

ROSSI: I would like to ask you a question. Recently some
             reports were published saying that patients with a
             positive sputum after treatment for over 40 days
             are to be regarded as non-infectious, even though
             some bacilli are still present in the sputum, they
             probably aren't virulent enough to infect their
             relatives or anyone who might come into contact
             with them. I would like to know your own opinion,
             whether or not we can trust these studies and
             insert such a patient again into social life.

SPINA: This depends on the possibility that isoniazide, a
             substance discovered by Middlebruch as long ago as
             1953 may reduce the virulence of the organisms,
             and make them no longer suitable for culture and
             therefore no longer alive. Middlebruch himself,
             however, pointed out, with a very appropriate
             sentence, that whilst it is true that isohiazide
             can make germs avirulent, it may be dangerous
             since these germs are likely to enter the cycle of
             infection, and we detect a primary resistence to
             isomiazide. I must admit, anyway, that this
             phenomenon occurs also with Rifampicin and
             therefore if it is true that in many cases these
             bacilli are seen, that is bacilli which are seen
             at a direct inspection, but not in culture or when
             injected into an experimental animal, they are
             always to be considered potentially dangerous.

VARADESE: I'd like to ask Professor Spina whether he has got
             any experience with pleural effusions during the
             spaced-out, bi-weekly treatment with Rifampicin or
             Isoniazide. This subject was discussed in Rome in
             1977 and we at the Campobasso Provincial
             Consortium have recorded two cases.

SPINA: This possibility was indeed considered, but if the
             therapy is correct, it hardly occurs: we haven't
             observed any such case during the spaced-out
             therapy we practised. The phenomenon however does
             exist, if the spaced-out therapy does not result
             in recovery or improvement. In this case a blood

dissemination or local spread to the pleura are possible, but that's a sign of ineffective therapy.

QUESTION: Among the parameters to assess the effectiveness of an antitubercular drug you listed serum concentrations as opposed to minimum inhibiting concentrations. I do not agree with you and I think you made two conflicting statements.

SPINA: Yes, it is clear that one must distinguish between serum concentration and concentration reached with drugs. Your objection anyway is justified by the revival experienced by perizinamide, which had been discarded just because it displayed its greatest effectiveness at the lesion site. As far as the ratio of the minimum inhibiting concentration to serum concentration is concerned, it's useful when it comes to assess the effectiveness of drugs, since, the higher this ratio is, the more effective against bacteria the drug is and the more likely it is to reach the lesion site in a concentration great enough to be effective.

DRUGS IN THE TREATMENT OF PRIMARY PULMONARY

ARTERIAL HYPERTENSION

C. Marini, G. Di Ricco, C. Guintini

C. N. R. Institute of Clinical Physiology and

2nd Medical Clinic, University of Pisa, Pisa, Italy

Introduction

Primary pulmonary arterial hypertension (PPH) should be suspected whenever the following features are present:

1 – history of progressive exertional dyspnoea, fatigue and repeated syncopal attacks especially in by young women.

2 – physical signs of pulmonary hypertension (increase of intensity of cardiac second sound) often associated with signs of low cardiac output (increase of intensity of cardiac first sound).

3 – ECG changes consistent with right ventricular hypertrophy.

4 – X-ray evidence of enlargement of right heart cavities and main pulmonary artery with normal left atrium.

5 – normal or only slightly altered pulmonary function.

Figure 1 shows the chest x-ray of a young woman with PPH. The right descending pulmonary artery has acquired a curved shape and has increased the usual distance from the border of the right atrium. The second arch of left cardiac border appears markedly prominent and it hides the left pulmonary artery. The angiogram (Figure 2) shows that this prominence is ascribable to the dilatation of the main pulmonary artery; both left and right branches appear to have a normal or slightly incremented size. Perfusion lung scan usually shows a base to apex gradient of pulmonary blood flow which is still similar to that normally seen.

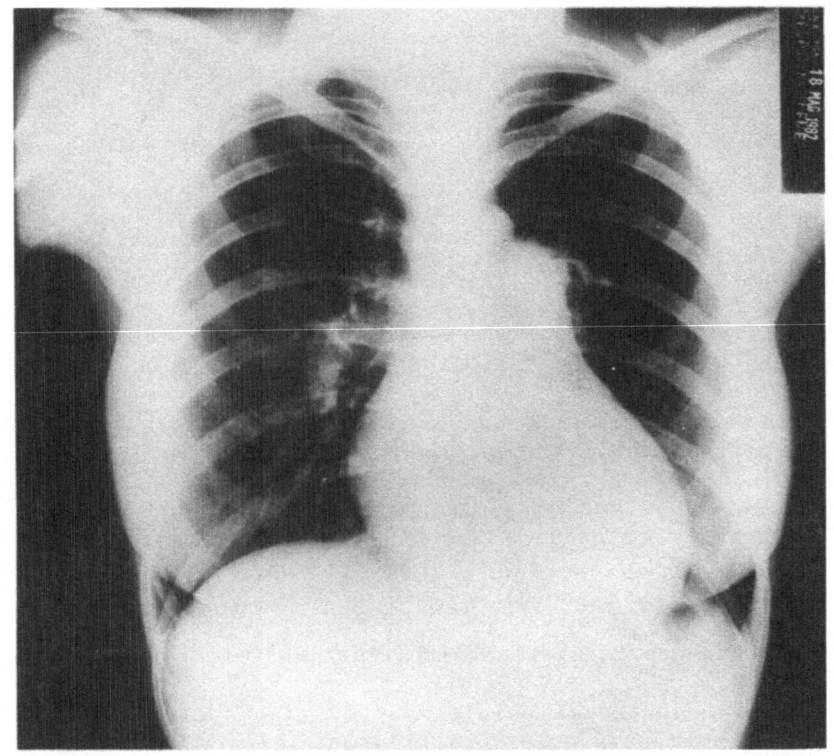

FIGURE 1:   Chest roentgenogram of a young woman with primary
              pulmonary arterial hypertension (see text).

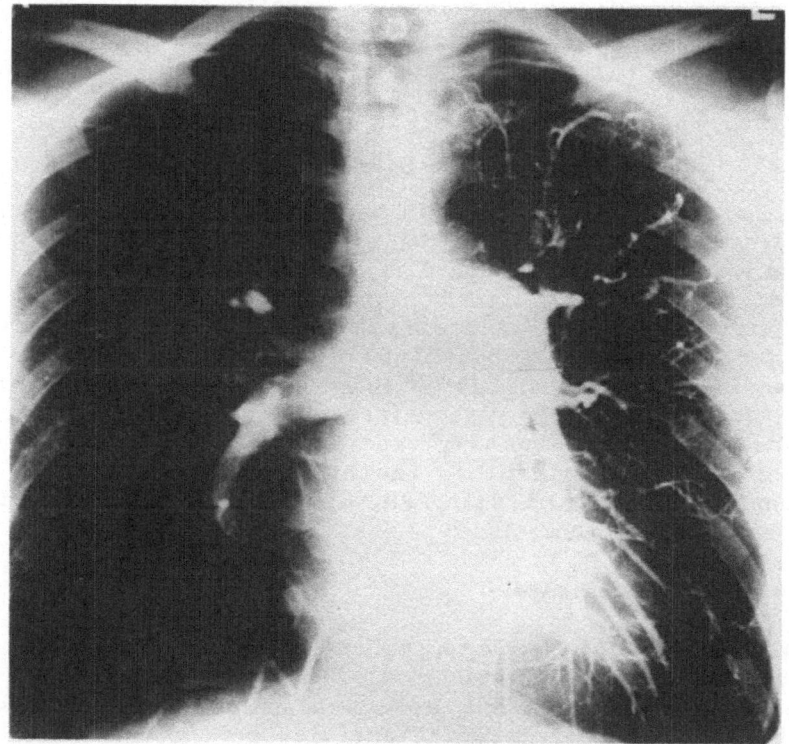

FIGURE 2:   Early phase of pulmonary angiogram performed in a woman with primary pulmonary arterial hypertension. Dilatation of the main pulmonary artery is evident.

The haemodynamic and angiographic findings that support the diagnosis of PPH are represented by:

a – increase of pulmonary artery pressure.

b – normal pulmonary wedge pressure.

c – absence of intracardiac or intrapulmonary shunts.

d – no evidence of pulmonary thromboembolic and veno-occlusive disease.

Because of these haemodynamic peculiarities it has been suggested that vasoconstriction of the pulmonary muscular arteries and arterioles plays a basic role in this disease. Hence PPH has become a clinical model to test drugs active on the pulmonary circulation. In this connection, appropriate haemodynamic criteria should be used to assess the favourable and untoward effects of each drug on the pulmonary circulation in PPH. Decrease of pulmonary artery pressure (PAP) together with pulmonary vascular resistances (PVR) and simultaneous increase of cardiac output (CO) during acutely performed pharmacological tests should be observed in order to prove both the extent of pulmonary vasoconstriction and the physiological responsiveness of pulmonary vasculature to a drug. In this respect, it seems useful to review the effects of drugs on pulmonary haemodynamics in PPH, assessed by means of right heart catheterisation.

Review of the Literature

The beta-agonist drugs, namely terbutaline and isoproterenol, have been tested in PPH by many authors (1 – 8) (Table 1). The prevalent effect of these drugs has been ascribed to their beta-mimetic activity on the heart. On the other hand, a favourable effect on the pulmonary circulation has been reported in 5 out of 19 patients reviewed in the literature (3 – 6). It seems noteworthy that, in the report by Person and Proctor (4), terbutaline was considered ineffective while isoprenaline in the same patient demonstrated a favourable effect on the pulmonary circulation.

The alpha-antagonists are another series of drugs tested in patients with PPH (Table II). Tolazoline, perhaps the oldest drug tried in this disease (2, 8 – 13) showed favourable effects on pulmonary vasculature in 8 out of 17 reported patients. As far as phentolamine is concerned, this drug

TABLE I

HEMODYNAMIC EFFECT OF DRUGS IN PPH: REVIEW OF THE LITERATURE

Beta - agonists

1) Isoproterenol

AUTHORS	REFERENCES		N. PATIENTS	EFFECTS	
D. Lee et al.	Am. Heart J.	1963	3	↓PVR; = PAP, PAw;	↑HR, PBV
BN. Rao et al.	Circulation	1969	1+	↓PAP,PVR;	↑HR, CO
UR. Shettigar et al.	N. Engl.J.Med.	1976	1x	↓PAP,PVR; = PAw;	↑HR, CO
B. Person and R.J. Proctor	Chest	1979	1x	↓PAP,PVR;	↑CO, PAw
JA. Pantano	N. Engl.J.Med.	1980	1	↓PAP,PVR	
E. Iupi-Herrera et al.	Chest	1981	6	↓PAP,PVR	↑HR, CO
			(no changes in 2)		
L. Gould et al.	Am. Heart J.	1981	1		↑HR, CO, PAP
S. Rich et al.	Abstract, Wien	1982	5°	↓PVR; = PAP	↑HR, CO

2) Terbutaline

B. Person et al.	Chest	1979	1x	Ineffective
		Tot.	19	

x same patient; + patient treated also with Tolazoline; ° series treated also with Tolazoline and Phentolamine

TABLE II

HEMODYNAMIC EFFECTS OF DRUGS IN PPH: REVIEW OF THE LITERATURE

Alpha - antagonists

1) Tolazoline

AUTHORS	REFERENCES	N. PATIENTS	EFFECTS
D. Dresdale et al.	Am. J. Med. , 1951	2	↓PAP,PVR
G. Wade et al.	Quart. J. Med. , 1957	1	Ineffective
P. Yu	Ann.Intern.Med. , 1958	6	↓PAP,PVR
R. Gorlin et al.	Br. Heart J. , 1958	1+	Ineffective
BN. Rao et al.	Circulation , 1969	1+	↓PVR; = PAP; ↑HR,CO
W. Fennel et al.	Chest , 1982	1	↓PVR; ↑HR,CO,PAP
S. Rich et al.	Abstract, Wien , 1982	5°	↓PVR; = PAP

2) Phentolamine

AUTHORS	REFERENCES	N. PATIENTS	EFFECTS
JN. Ruskin et al.	Ann.Intern.Med. , 1979	1	↓PAP,PVR; = PAw; ↑CO
Se Do Cha et al.	Ann.Intern.Med. , 1979	1	Ineffective
HL. Cohen et al.	Ann.Intern.Med. , 1981	1	↑HR,CO,PAP,PVR
L. Gould et al.	Am. Heart J. , 1981	3 (in 2)	↓PAP,PVR. ; ↑CO
S. Rich et al.	Abstract, Wien , 1982	5°	↓PVR; = PAP
	Tot.	23	

+ patient treated also with isoproterenol; ° series treated also with isoproterenol.

showed a favourable effect in 3 patients (7, 14) while was ineffective in two (7, 15). In one additional patient (16) the effect of this drug was considered as dependent upon beta stimulation of the heart. Finally, in the series of Rich et al. (8), the effect of phentolamine was considered inconsistent to produce effective pulmonary vasodilatation. In summary, a favourable effect of alpha-antagonist drugs has been demonstrated in 11 out of 23 patients.

Calcium-antagonists have also been tried in PPH (Table III). As far as verapamil is concerned, in the series of Landmark et al. (17), a favourable effect was obtained in 2 out of 9 patients with PPH. In the other 7 patients, verapamil led to either a drastic decrease in cardiac output or to a marked strain on the right ventricle. Nifedipine showed, in one case reported by Camerini et al (18), an effective pulmonary vasodilatation, while in the series of Rozkovec et al (20), while in 4 patients reported by Crevey et al (21) data rather suggest a remarkable vasodilatation on the systemic circulation.

Some other drugs have been tried in patients with PPH (Table IV). Hydrallazine, in the series of Rubin et al. (22), showed an effect of vasodilatation on the pulmonary circulation. However, these authors suggest that the effect may be due to an indirect effect on pulmonary circulation via baroreceptors. Lupi-Herrera et al (23) reported 12 cases of PPH treated with hydrallazine: in 6 of them a marked reduction of right heart afterload was obtained, while in the remaining 6 the main effect of the drug was a remarkable systemic vasodilatation. Prostacyclin did not show any selective effect on the pulmonary circulation (19, 24). Diazoxide showed a favourable effect, at right heart catheterisation, in one patient during oral treatment (29) and in one patient reported by Wang et al (25). In 7 out of 9 patients of the series described by Honey et al (27), diazoxide was effective in increasing pulmonary blood volume. PAP fell by more than 10 mmHg in only two patients. Unfortunately, in this series, diazoxide showed side effects such as nausea and sickness, diabetes, peripheral oedema, and postural hypotension in 5 out of 7 patients put on oral diazoxide so that these 5 patients stopped the treatment. Two patients showed sustained clinical improvement while remaining on treatment but developed troublesome hirsutes.

As far as other reports on diazoxide are concerned, the drug showed a marked systemic vasodilatation (26, 28) and some effects on atrio-ventricular conduction up to A-V block (30). Finally, captopril, a drug that blocks the angiotensin converting enyzme, did not show any effect on the pulmonary

TABLE III

HEMODYNAMIC EFFECTS OF DRUGS IN PPH: REVIEW OF LITERATURE

Calcium antagonists

1) Verapamil

AUTHORS	REFERENCES	N. PATIENTS	EFFECTS	
K. Landmark et al.	Acta Med. Scand., 1978	9 (in 2)	↓PAP;PVR	↑CO

2) Nifedipine

F. Camerini et al.	Br. Heart J. , 1980	1*	↓PAP,PVR;	↑HR,CO
A. Rozkovec et al.	Abstract, Wien 1982	6+(in 2)	↓PVR	

3) Diltiazem

H. Kambara et al.	Am. Heart J. , 1981	1	↓PAP,PVR;	↑CO
BJ. Crevey et al.	Am. J. Cardiol. , 1982	4	↓PAP,PVR,BP;	= CO
	Tot.	21		

* patient treated also with diazoxide; + series treated also with PIG$_2$

TABLE IV

HEMODYNAMIC EFFECTS OF DRUGS IN PPH: REVIEW OF THE LITERATURE

1) Hydralazine

AUTHORS	REFERENCES	N. PATIENTS	EFFECTS	Others
LJ. Rubin et al.	N. Engl. J. Med., 1980	4	↓PVR; = PAP;	↑HR,CO
E. Lupi- Herrera et al.	Circulation , 1982	12 (in 6)	↓PVR; = PAP;	↑HR,CO

2) Prostacyclin (PGI_2)

AUTHORS	REFERENCES	N. PATIENTS	EFFECTS	Others
DN. Guadagni et al.	Br. Heart J. , 1981	4	Ineffective	
A. Rozkovec et al.	Abstract, Wien 1982	6[+](in 2)	↓PVR	

3) Diazoxide

AUTHORS	REFERENCES	N. PATIENTS	EFFECTS	Others
SWS. Wang et al.	Br. Heart J. , 1978	3	↓PVR, BP;	↑HR, CO
JM. Rubino et al.	Br. Heart J. , 1979	1 up to 180 mg with 300 mg ↓BP, shock and death	= CO;	↑HR,PAP
M. Honey et al.	Thorax , 1980	9	↓PVR,BP; = PAP;	↑PBV
WP. Klinke et al.	N. Engl. J. Med., 1980	1*	↓PAP,PVR, BP;	↑HR,CO
F. Camerini et al.	Br. Heart J. , 1980	1	↓PVR, BP; = PAP	↑CO
DR. Hall et al.	Br. Med. J. , 1981	1 (oral therapy)	↓PAP,PVR;	↑CO
J. Buch et al.	Br. Heart J. , 1981	3	Important side effects (AV block, marked hypotension)	

4) Captopril (ACE - blocker)

AUTHORS	REFERENCES	N. PATIENTS	EFFECTS	Others
S. Rich et al.	Abstract, Wien 1982	4	Ineffective	
	Tot.	49		

* patient treated also with nifedipine; + series treated also with nifedipine

circulation (31). Some other drugs may be tried in PPH on the basis of their vasodilator effect demonstrated in pulmonary hypertension of different origin, or in experimental animals (Table V).

Prazosin could be used in PPH instead of phentolamine for long term treatment in the attempt to reduce the side effects described in the prolonged use of phentolamine.

It has been suggested that pharmacological effects on pulmonary circulation in PPH should be more evident when the disease is at an early stage, namely when active vasoconstriction seems to be the dominant element of the disease. However, morphological studies have shown that irreversible lesions such as angionecrosis, sometimes considered as a consequence of prolonged vasoconstriction, may develop independently from vasoconstriction itself (32). It appears, then, that the effect of vasoactive drugs cannot be correlated with the stage of PPH.

MATERIALS AND METHODS

The haemodynamic resting data concerning 7 patients with PPH of our series are shown in table VI. All patients were females: mean age was 35.7 years (range from 13 to 45 years). Right atrial pressure (RAP, mean 7.0 mmHg) was within the normal range of our laboratory in 4 patients. In patients No. 2, 6 and 7 RAP was elevated and was associated tricuspid with incompetence. Mean pulmonary artery pressure (MPAP) was markedly raised in all patients (mean 50.1 mmHg) and so were pulmonary vascular resistances (PVR, mean 20.0 mmHg/(L/min/M^2) and total pulmonary vascular resistances (TPVR, mean 23.3 mmHg/(L/min/M^2). Cardiac Index (CI) was low in most patients (mean 2.64 L/min/M^2), ranging from 1.4 to 4.4 Lmin/M^2.

Table VII shows the list of vasoactive drugs acutely tested during right heart catheterisation and, for each drug, the dose administered. We performed 20 pharmacological tests in 7 patients. The values of the parameters used in this report are those recorded at the time of maximum haemodynamic effect of each drug.

Four out of our 7 patients performed 100% O_2 inhalation for 15 minutes (Figure 3) in order to assess the role of hypoxemia in this disease. One of our patients, namely that represented with a filled circle in Figure 3, was reported as having hypoxic pulmonary hypertension by a highly experienced pathologist who had examined her lung biopsy. One hundred per cent O_2 did not change significantly MPAP in the patients

MPAP VARIATION IN PPH BEFORE[B] AND AFTER[A]
O$_2$ 100% ADMINISTRATION FOR 15 MINUTES

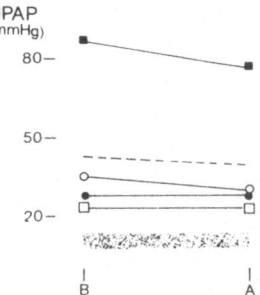

FIGURE 3:   Effect of 100% O$_2$ inhalation on mean pulmonary
            arterial pressure (MPAP).
            The shaded area represents the normal range and
            the dashed line the mean values of the 4 patients
            before (B) and during (A) oxygen inhalation.

HEMODYNAMICS IN PPH BEFORE (B) AND AFTER(A) ISOPROTERENOL

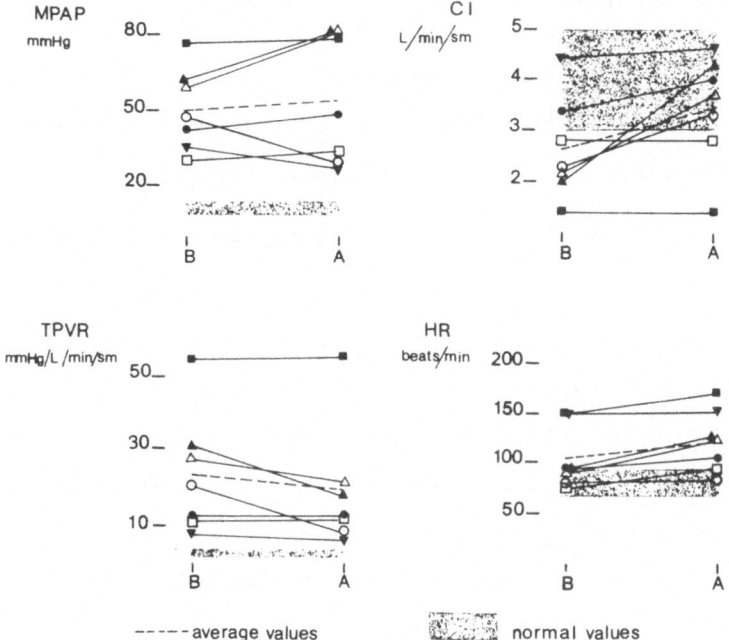

FIGURE 4:   Haemodynamic changes following isoproterenol.
            MPAP: mean pulmonary artery pressure: CI: cardiac
            index.   TPVR: total   pulmonary   vascular
            resistance; HR: heart rate.

TABLE V

ADDITIONAL DRUGS THAT MAY BE TRIED IN PPH

DRUGS	MAIN MODE OF ACTION
Aminophylline	(vasodilator)
Amyl nitrite	"
Isosorbide dinitrate	"
Prazosin	(alpha antagonist)
Indomethacin	(prostaglandin antagonist)

TABLE VI

HEMODYNAMIC DATA IN 7 PATIENTS WITH PPH

	INITIALS	AGE years	SEX	RAP mmHg	PAP(Mean) mmHg	MPWP mmHg	CI $L/min/M^2$	PVR $\frac{mmHg}{L/min/M^2}$	TPVR $\frac{mmHg}{L/min/M^2}$
1	CB	36	F	5	72/31(47)	8	2.3	17.0	20.4
2	MM	40	F	11	93/56(76)	8	1.4	48.6	54.3
3	SL	13	F	0	45/20(35)	1	4.4	7.7	7.9
4	LB	43	F	1	50/15(30)	1	2.8	10.4	10.7
5	IA	32	F	4	67/29(42)	8	3.4	10.0	12.3
6	GG	38	F	15	90/44(60)	12	2.2	21.8	27.2
7	MA	45	F.	13	95/42(61)	12	2.0	24.5	30.5
	MEAN	35.7		7.0	(50.1)	7.1	2.64	20.0	23.3

TABLE VII

STUDY PROTOCOL

DRUGS		DOSE (mg)	WAY OF ADMINISTRATION	N° OF PATIENTS
Isoproterenol	(I)	20	sublingually	7
Phentolamine	(P)	5	i v	3
Diazoxide	(D)	270-450	i v	2
Aminophylline	(A)	200	i v	4
Amyl Nitrite	(N)	*	inhalation	4

* To double resting heart rate

TABLE VIII

EFFECTS OF DRUGS ON PULMONARY HEMODYNAMIC
IN 7 PATIENTS WITH PPH

DRUGS	No of patients	MPAP mmHg		TPVR $\frac{mmHg}{L/min/M^2}$		CI $L/min/M^2$	
		Average change	Range	Average change	Range	Average change	Range
Isoproterenol	7	+3.6	-18/+20	-4.0	-11.5/+1.4	+0.78	0/+2.2
Phentolamine	3	-3.7	- 8/+4	-4.0	- 7.2/-2.3	+0.33	+0.2/+0.4
Diazoxide °	2	+7.5	+ 4/+11	-0.9	- 1.8/-0.1	+0.70	+0.4/+1.0
Aminophylline	4	-2.8	- 8/+1	-2.3	- 4.2/-0.4	+0.20	0/+0.3
Amyl nitrite	4	+1.7	0/+4	-1.0	- 5.3/+1.0	+0.20	0/+0.7

° values recorded before side effects

Increments and decrements of hemodynamic parameters are in units of measure not as percent.

including the one represented with a filled circle. This
particular case raises serious doubts on the ability of
morphologic criteria to classify patients with pulmonary
hypertension. More generally, these findings support the
belief that hypoxia does not play an important role in
determining the extent of pulmonary vasoconstriction in these
patients.

RESULTS

Figure 4 shows the changes of CI, MPAP, TPVR and heart rate
(HR) after isoproterenol in the 7 patients of our series (see
also table VIII). Whereas CI and HR increased, TPVR tended to
decrease, and MPAP behaved differently among the patients.
Namely, in 3 patients (filled circle, filled and open
triangles) MPAP increased by 6, 19 and 20 mmHg, respectively,
whereas in two decreased and in the other two did not change
significantly. On the whole, in our 7 patients after
isoproterenol, the mean values of MPAP resulted higher than
that at rest.

Figure 5 shows the haemodynamic changes after phentolamine
in 3 patients (see also Table VIII). It is noteworthy that 2
of the 3 patients, who showed MPAP increase after
isoproterenol, displayed a decrease of MPAP after phentolamine.
On the other hand, since CI slightly increased, TPVR on the
average slightly decreased in all the 3 patients. The HR did
not show significant changes.

Diazoxide, as reported in the literature (26, 30), when
administered in the pulmonary artery of our patients, produced
major side effects such as shock, nausea, and vomiting. From
the haemodynamic point of view (Table VIII), it appears
difficult to interpret the effect of diazoxide on the pulmonary
circulation in our two patients since the haemodynamic changes
appeared markedly influences by the vasodilatation effect on
the systemic circulation.

Haemodynamic changes after aminophylline were similar to
those observed after phentolamine but of minor extent (Table
VIII).

After amyl nitrate no significant haemodynamic changes were
observed with the dosage used.

DISCUSSION

Our data, as far as the effect of isoprenaline is concerned,
suggest that the haemodynamic changes induced by this drug may

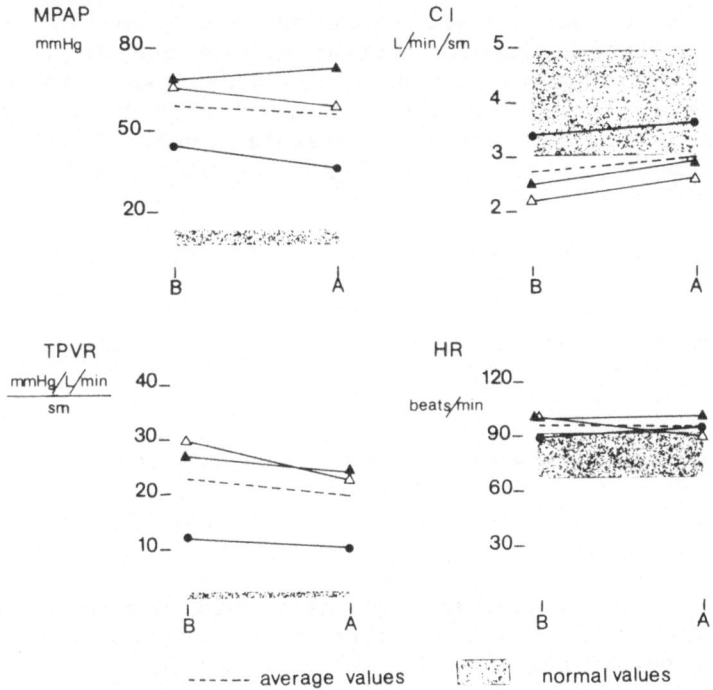

FIGURE 5:   Haemodynamic changes induced by phentolamine.
MPAP: mean pulmonary artery pressure; CI: cardiac
index; TPVR: total pulmonary vascular resistance;
HR: heart rate.

be mainly ascribed to a beta-mimetic effect on the heart. On
the other hand, after phentolamine, the same fall of TPVR is
accompanied by an average slight decrease of MPAP, while CI is
slightly increased and HR is unchanged. In this case, then,
the data suggest a specific effect of phentolamine on alpha-
receptors in the pulmonary vasculature leading to pulmonary
vasodilatation by an alpha-lytic mechanism. This is in
agreement with the physiological knowledge on the distribution
of alpha-receptors in the pulmonary arteries (33) and the
experimental data reported by Zener and Harrison (34)
concerning the pharmacological effect of phentolamine. Just
before the dose which caused the untoward side effects in the
two patients of our series, diazoxide showed roughly the same
effects as isoprenaline. Aminophylline and amyl nitrite
influenced slightly the pulmonary haemodynamics. Aminophylline
effect, even though of minor relevance in PPH, might be
explained by a direct action on vascular smooth muscle on the
human pulmonary circulation (35). On the other hand, amyl
nitrite practically increased only HR.

In summary, the effect of drugs on PPH can be described as
follows:

1. Patients may have a widely different response to a given
   drug.

2. A patient may respond differently to different drugs:
   alpha-antagonists seem more effective than beta-agonists.

3. The haemodynamic response cannot be used to predict the
   stage of PPH.

4. The effect is considered favourable when pulmonary arterial
   pressures fall together with vascular resistances and
   cardiac output rises.

5. This favourable effect is rare in patients with PPH.

In conclusion, right heart catheterisation should be
performed on patients with PPH to evaluate pulmonary vascular
responsiveness to vasoactive drugs. Occasional favourable
responses have been reported with several drugs, but the
implications of such acute responses on the long term treatment
with the drugs proved acutely effective have yet to be
established.

REFERENCES

1. Lee, T. D. Jr., Roveti, G. C., Ross, R. S. 1963.
   The hemodynamic effects of Isoproterenol on pulmonary
   hypertension in man.
   Am. Heart J. 65 (3), 361-367.

2. Rao, B. N. S., Moller, J. H. and Edwards, J. E. 1969.
   Primary pulmonary hypertension in a child. Response to
   pharmacologic agents.
   Circulation XL: 583-587.

3. Shettigar, U. R., Hultgren, H. N., Specter, M., Martin, R.
   and Davies, D. H. 1976.
   Primary pulmonary hypertension. Favourable effect of
   Isoproterenol.
   N. Engl. J. Med., 295 (25), 1414-1415.

4. Person, B. and Proctor, R. J. 1979.
   Primary pulmonary hypertension. Response to
   Indomethacin, Terbutaline and Isoproterenol.
   Chest, 76 (5), 601-603.

5. Pantano, J. A. 1980.
   Isoproterenol in primary pulmonary hypertension.
   N. Engl. J. Med., 302 (16), 919-920.

6. Lupi-Herrera, E., Bialostozki, D. and Sobrino, A.
   The role of Isoproterenol in pulmonary artery
   hypertension of unknown etiology (primary). Short and
   long-term evaluation.
   Chest, 79 (3), 292-296.

7. Gould, L., Chokshi, A. B., Patel, S., Gomes, G. I. 1981.
   Hemodynamic evaluation of vasodilator drug therapy in
   primary pulmonary hypertension.
   Am. Heart J., 102 (2), 300.

8. Rich, S., Levy, P. S., Rosen, K. M. 1982.
   A reassessment of the acute hemodynamic effects of
   vasodilators in primary pulmonary hypertension.
   Symposium on primary pulmonary hypertension.
   Wien, 18-20 March, Abstract book, page 39.

9.   Dresdale, D. T., Schultz, M. and Michtom, R. J. 1951.
     Primary pulmonary hypertension: I. Clinical and
     hemodynamic study.
     Am. J. Med., 11, 686.

10. Wade, G. and Ball, J. 1957.
     Unexplained pulmonary hypertension.
     Quart. J. Med., 26, 83.

11. Yu, P. N. 1958.
     Primary pulmonary hypertension: report of six cases and
     review of literature.
     Ann. Intern. Med., 49 (5), 1138-1159.

12. Gorlin, R., Clare, F. B., Zuska, J. J. 1958.
     Evidence of pulmonary vaso-constriction in man.
     Br. Heart J., 20, 346.

13. Fennell, W. H., Farmer, J. A. Graf, R. H. and Young, J. B.
     1982.
     Unanticipated response to alpha-adrenergic blockade in
     pulmonary hypertension.
     Chest, 81 (1), 128-129.

14. Ruskin, J. N. and Hutter, A. M. 1979.
     Primary pulmonary hypertension treated with oral
     phentolamine.
     Ann. Intern. Med., 90, 772-774.

15. Se Do Cha., Kirschbaum, M., Maranhao, V., Paine, E.,
     Gooch, A. S. 1979.
     Phentolamine for primary pulmonary hypertension.
     Ann. Intern. Med., 91 (6), 927-928.

16. Cohen, M. L. and Kronzon, I. 1981.
     Adverse hemodynamic effects of Phentolamine in primary
     pulmonary hypertension.
     Ann. Intern. Med., 95 (5), 591-592.

17. Landmark, K., Refsum, A. M., Simonsen, S. and Storstein,
     O. 1978.
     Verapamil and pulmonary hypertension.
     Acta Med. Scan., 204, 299-302.

18. Camerini, F., Alberti, E., Klugmann, S., Salvi, A. 1980.
     Primary pulmonary hypertension: effects of nifedipine.
     Br. Heart J. 44, 352-356.

19. Rozkovec, A., Stradling, J., Shepherd, G., Wilde, S.,
     MacDermot, J., Oakley, C. N. and Dollery, C. T. 1982.
     Variable response to vasodilator therapy in primary
     pulmonary hypertension.
     Symposium on primary pulmonary hypertension, Wien, 18-
     20 March. Abstracts book page 38.

20. Kambara, H., Fujimoto, K., Wakabayashi, A and Kawai, C.
     1981.
     Primary pulmonary hypertension: beneficial therapy with
     diltiazem.
     Am. Heart J., 101 (2), 230-231.

21. Crevey, B. J., Dantzker, D. R., Bower, J. S., Popat, K.
     D., Walker, S. D. 1982.
     Hemodynamic and gas exchange effects of intravenous
     Diltiazem in patients with pulmonary hypertension.
     Am. J. Cardiol., 49, 578-583.

22. Rubin, L. J. and Peter, R. H. 1980.
     Oral hydralazine therapy for primary pulmonary
     hypertension.
     N. Engl. J. Med., 302 (2), 69-73.

23. Lupi-Herrera, E., Sandoval, J., Seoane, M. and
     Bialostozky, D. 1982.
     The role of Hydralazine therapy for pulmonary arterial
     hypertension of unknown cause.
     Circulation, 65 (4), 645-650.

24. Guadagni, D. N., Ikram, H., Maslowski, A. H. 1981.
     Haemodynamic effects of prostacyclin (PGI_2) in
     pulmonary hypertension.
     Br. Heart J., 45, 385-388.

25. Wang, S. W. S., Pohl., J.E.F., Rowlands, D. J. and Wade,
     G. 1978.
     Diazoxide in treatment of primary pulmonary
     hypertension.
     Br. Heart J., 40, 572-574.

26. Rubino, J. M. and Schroeder, J. S. 1979.
     Diazoxide in treatment of primary pulmonary
     hypertension.
     Br. Heart J., 42, 362-363.

27. Honey, M., Cotter, L., Davies, N. and Denison, D. 1980.
     Clinical and haemodynamic effects of diazoxide in
     primary pulmonary hypertension.
     Thorax, 35, 269–276.

28. Klinke, W.P. and Gilbert, J. A. L. 1980.
     Diazoxide in primary pulmonary hypertension.
     N. Engl. J. Med., 302 (2), 91–92.

29. Hall, D. R., Petch, M. C. 1981.
     Remission of primary pulmonary hypertension during
     treatment with diazoxide.
     Br. Heart J., 282, 1118.

30. Buch, J., Wennevold, A. 1981.
     Hazards of diazoxide in pulmonary hypertension.
     Br. Heart J., 46, 401–403.

31. Rich, S., Rosen, K. M. 1982.
     Captopril for primary pulmonary hypertension: the
     problem of variability in assessing chronic drug
     therapy.
     Symposium on primary pulmonary hypertension.
     Wien, 18–20 March. Abstracts book, page 29.

32. Ogata, T. and Watanabe, S. 1982.
     Pulmonary vascular changes in 32 autopsy cases with
     unexplained pulmonary hypertension in Japan.
     Symposium on primary pulmonary hypertension.
     Wien, 18–20 March. Abstracts book, page 4.

33. Bergofsky, E. H. 1980.
     Humoral control of the pulmonary circulation.
     Ann. Rev. Physiol., 42, 221–233.

34. Zener, J and Harrison, D. C. 1973.
     Circulatory actions of Phentolamine before and after
     autonomic blockade.
     Cardiov. Res. 7, 748–754.

35. Harris, P., Heat, D. 1977.
     The Human Pulmonary Circulation. 2nd Edition.
     Churchill Livingstone, Edinburgh, page 200.

DISCUSSION

SPEAKER: MARINI   **CHAIRMAN: CUMMING**

CHAIRMAN: This paper is now open for discussion.

DENISON: Dr. Marini, when you began you said that primary
                 pulmonary hypertension was frequently used as a
                 model of pulmonary hypertension generally. You
                 showed, as many other people showed, that in other
                 forms of pulmonary hypertension you get quite a
                 marked reversal with oxygen, but in primary
                 pulmonary hypertension you didn't and it seems
                 that there is very little response to any of the
                 drugs that you tried. So my first question is: do
                 you still think that it should be used as a model
                 of pulmonary hypertension generally or should it
                 be abandoned because it is so resistant to any
                 sort of intervention? I have got two other
                 questions but perhaps you'd like to answer that
                 first.

MARINI: Thank you Dr. Denison for this question. Let us
                 say that the etiology of primary pulmonary
                 hypertension is still virtually unknown. The role
                 vasoconstriction may play in this disease has been
                 much talked about for a long time but, I should
                 say that as long ago as 1958, that is when Yu
                 first described the disease on the basis of his
                 own observations, it already started to become
                 clearer that the role vasoconstriction played in
                 this disease was lower when compared with the
                 other pathologies and more specifically a
                 pathology which affects pulmonary vessels more
                 directly.

DENISON: There was a curious phrase that you used in which
                 you said that there had been a fall in pulmonary
                 vascular resistance but this need not have been at
                 a pulmonary level. I am not sure what you meant
                 by it.

MARINI: According to the way we regard changes in
                 pulmonary haemodynamics occurring after

vasodilation, a decreased resistance should always involve a decreased pulmonary pressure. If that's not the case, it doesn't mean that there has been no vasodilation, but that its effects are not blatant, if pulmonary pressure remains unchanged.

DENISON: This leads directly to my third question in that case. In one of the papers you mentioned by Honey, we showed there that when you gave diazoxide the pulmonary vascular resistance, calculated from the pressure drop fell quite markedly in some people by twofold, the pulmonary arterial pressure remains constant because cardiac output would have risen, sometimes 2.1/2 times as much. It's his interpretation of these findings that I find interesting since it implies that the pulmonary vascular bed is capable of dilating, on the other hand, if you calcuate the work of the right heart, since the pressure drop across it is the same and the flow has gone up, you've actually increased the work of the heart and that's probably a bad thing, so how do you interpret this, is it a good thing to produce pulmonary dilatation if it is not accompanied by a fall in pressure? The point where you failed to answer my question is: if you calculate the vascular resistance and it has fallen, it doesn't imply that there is a reversable element.

MARINI: To answer your first question I should say that I agree with you that if median pulmonary pressure is not reduced, the work load on the right ventricle increases and this as a result could entail disadvantages, for instance to cause right cardiac decompensation. As to your second question, I can't rule out the possibility that a reduction in pulmonary vascular resistance has been obtained, as stated in Honey's work, and therefore it is assumed, also because pulmonary blood volume increased at the same time, that there has been vaso dilation; what I meant is that this occurred in two patients only out of nine.

VEGETATIVE CONTROL OF PULMONARY VASCULAR COMPLIANCE

AND RESISTANCE IN MAN

Carlo Guintini

C.N.R. Institute of Clinical Physiology and 2nd Medical

Clinic of the University of Pisa, Pisa, Italy

The pulmonary vessels are controlled by mechanical factors, especially gravity and the respiratory movements of the lung, by the autonomic nervous system, by changing oxygen and hydrogen ion concentrations and by vasoactive substances. These factors have been studied in controlled animal experiments but we remain to a large extent ignorant as to how they interact in life and, in particular, in man.

1. Anatomical layout of the pulmonary circulation and its innervation in the adult human lung.

The adult pulmonary artery is considered a fairly distensible vessel even in recent publications (1). Intrapulmonary branches are muscular and accompany the branching airways down to the level of the respiratory bronchioles. More distal vessels are thin walled with a single elastic lamina and little or no muscle. They break up into alveolar capillaries which, according to different workers, form either a network (2) or a continuous sheet in the alveolar wall supported by pillars of connective tissue (3). The nature of capillary blood flow will differ in the two cases. Collecting venules are thin-walled and lead into veins with thin muscular coats which lie at the periphery of the lobules.

Knowledge on the distribution of the nerves to the pulmonary vessels is still incomplete in humans. It has been established that the innervation of the human pulmonary circulation is partly parasympathetic via the vagus and partly sympathetic; that it is both sensory and motor in function and that the nerve-fibres concerned are carried in several nerve-bundles that enter the lung at the hilum in company, for the most part, with the bronchi. The problem of definite morphologic identification of sensory endings in the lung circulation is unresolved, although the physiological evidence for their existance is firm (4). The pulmonary veins show a complex network of nerve fibres extending to the subendothelial region, which was considered to supply endings both baroceptive (deep) and chemoreceptive (superficial) in nature (5). In the alveolar walls, unmyelinated axons have been located in the interstitium between pneumocytes and the capillaries (6). They terminate with enlarged nerve endings suggested to be sensory in function (J or juxta-capillary-receptors). Sensory nerve cells have been located in the cervical and thoracic portions of the vagus nerve (7). These neurons were believed to be present in addition to the neurons in the inferior vagal ganglion or nodose ganglion (5). For this reason, afferent fibres are thought to run with the vagal branches. Efferent fibres supply both the pulmonary veins and arteries, derive from vagal and sympathetic branches, and constitute the motor innervation. The pulmonary arterial tree is innervated by an investing plexus of nerve fibres as far as the terminal arterioles and fibres ramify in the adventitia to form the primary plexus. This plexus of unmyelinated nerves contains numerous Schwann-cell nuclei. In continuation with this primary plexus, the terminal effector plexus is essentially limited to the adventitio-medial junction. There is focal superficial penetration of the media varying somewhat according to the calibre of the vessel. Nodal specilisations of axoplasms represent bare nerve terminal areas (8). Demonstration of the distribution of motor innervation to both pulmonary arteries and veins has been obtained by special techniques. Staining for acetylcholinesterase has been used to identify cholinergic nerves. Fluorescence induced with formaldehyde gas has been applied to identify adrenergic nerves (9). Experiments involving a variety of agonists and antagonists indicate that both alpha and beta adrenergic receptors are plentiful (10).

What is the function of the innervation of the vessels in the pulmonary circulation of a living human being? In general terms, an answer to this question is impossible on the basis of available data. I shall confine myself to present two examples relating principally to pulmonary vascular compliance and resistance.

2. Pulmonary vascular compliance

Since the infusion of a plasma expander raises both the

pulmonary arterial and wedge pressures to the same extent, such
an infusion provides an opportunity to assess the compliance of
the pulmonary vascular bed by determining the ratio of the
change in volume to the change in pressure (11). As shown in
Figure 1, the values for the distensibility of the pulmonary
circulation obtained following dextran infusion appear
exceedingly low. The pulmonary vascular compliance remains at
a low level for at least one hour after the infusion in the
absence of any intervention. Since the compliance values are
low, in presence of elevated pulmonary vascular pressures,
whereas pulmonary blood volume and also pulmonary blood flow,
(not shown in the Figure) present only minor increments, it may
be concluded that the concept of the pulmonary vascular bed as
a highly distensible system is not valid in these circumstances
and that an active mechanism is probably opposing the
distension of the pulmonary vessels in the face of an
increasing transmural pressure. An analysis of the behaviour
of the pulmonary vascular resistance shows a small decline of
pulmonary vascular resistance in the presence of a marked
elevation of mean pulmonary wedge pressure, Figure 2 thus
suggesting vasomotor activity in the resistance vessels
opposing their passive distension. These findings may be
tentatively interpreted by the hypothesis that acute elevation
of left heart and pulmonary venous pressures increases the
stiffness of the pulmonary vasculature, and the lung as a
whole, via a reflex mechanism. This might originate from
baroreceptors in the area of the left heart and large pulmonary
veins and have its efferent arm in sympathetic fibres. Indeed,
in the dog, sympathetic stimulation increases the stiffness of
the larger lobar arteries but has little effect on calculated
pulmonary vascular resistance (12). Phenoxybenzamine elicits
opposite responses, indicating that the effects are largely due
to stimulation of Alpha adrenergic receptors. Propranolol does
not blunt this increase in pulmonary vascular stiffening (13).
Hence, by altering the physical properties of the pulmonary
vessels, sympathetic activation could augment the hydraulic
load presented to the right ventricle (13). This, in turn,
might impair the output of the right ventricle thus preventing
development of pulmonary oedema in situations such as after
dextran infusion. This situation appears to persist for quite
some time. Thus, vegetative control of the pulmonary
circulation would not just be a vestigial function.

At this stage, it may be interesting to test some drug that
could enhance the distensibility of the pulmonary vascular bed.
Atropine sulphate, which is known to lower central venous
pressure (14), is found to produce a striking increase of the
compliance of the lung circulation when injected intravenously
in doses of 2 to 3 mg after the infusion of dextran (11). The

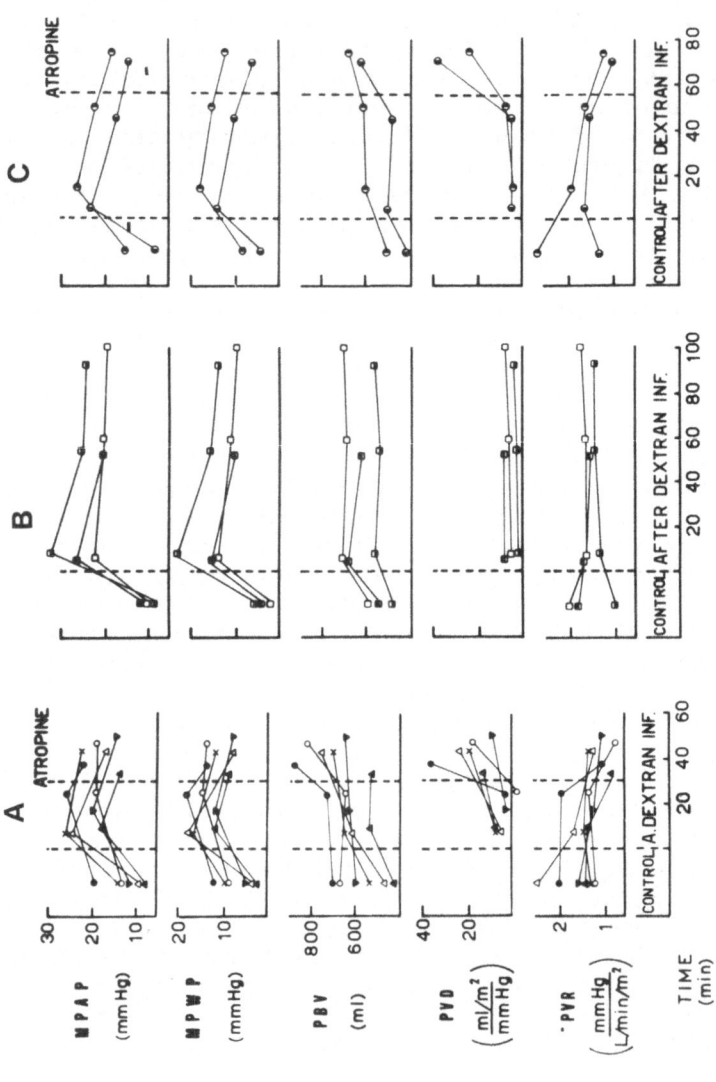

FIGURE 1:   Haemodynamic changes in the pulmonary circulation following dextran infusion and
subsequent atropine injection. Sections A, B and C report data on different
normal subjects. MPAP: mean pulmonary arterial pressure; MPWP: mean pulmonary
wedge pressure; PBV: pulmonary blood volume; PVD: pulmonary vascular
distensibility; PVR: pulmonary vascular resistance. Values are derived from
reference (11).

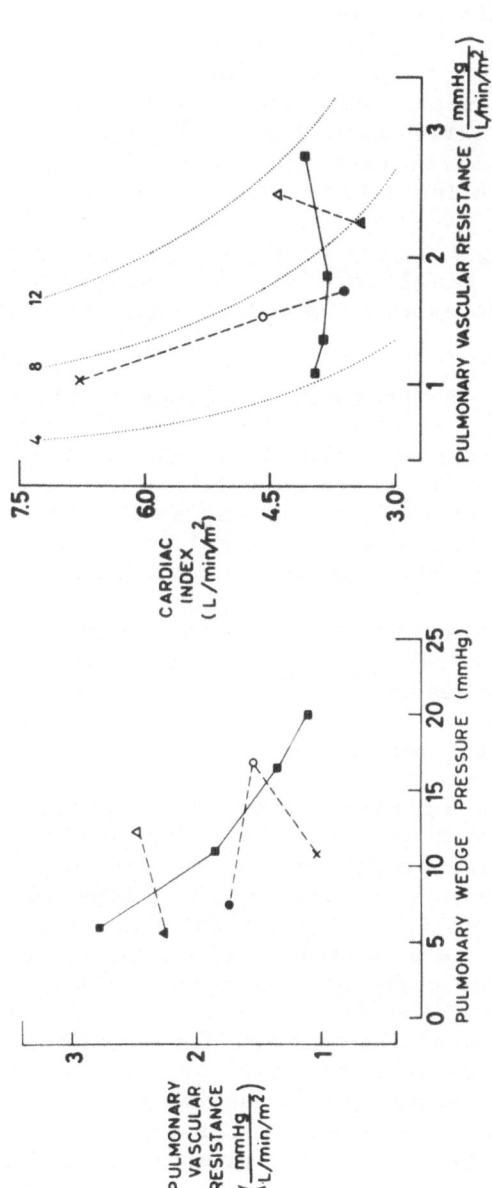

FIGURE 2: Pulmonary vascular resistance versus pulmonary wedge pressure (left hand panel) and cardiac index (right hand panel), respectively, during control (closed circle), following dextran infusion (open circle), and subsequent atropine injection (cross), in normal subjects. Other symbols as in reference (11).

changes of pulmonary vascular compliance and resistance after
the injection of atropine appear consonant with the hypothesis
of a reflex mechanism for the following reasons: -

(1) relaxation resulting from the parasympatholytic action of
atropine is probably important for the bronchial musculature,
but, on the basis of present knowledge (15), cannot be
postulated for the human lung vessels. Even a direct action of
atropine on pulmonary vessels appears unlikely, since no
evidence has been collected in man or dog that indicates
pulmonary vasodilatation after atropine administration (16-18).

(2) Relaxation of pulmonary vessels after atropine does not
seem ascribable to ganglionic blockade, since this action of
the drug requires doses much larger than those employed in our
study (19,20).

(3) On the other hand, after dextran infusion the stroke work
of the left ventricle is 65 gm per beat per m^2 with a LVEDP of
16 mmHg, whereas after atropine the stroke work is 67 g-m per
beat per m^2 with a LVEDP of 10 mmHg (21). This enhancement of
the left ventricular performance may be interpreted, on the
basis of some evidence of a tonic negative inotropic influence
of the vagus nerve upon mammalian ventricular myocardium
(22,23), as the result of the parasympatholytic activity of the
drug on the ventricle, and it may be considered at the origin
of the pressure decrease which, in turn, may interrupt the
reflex mechanism suggested before.

3. Pulmonary vascular resistance

Several authors criticise the physiologic meaning of
computed pulmonary vascular resistance and postulate the
necessity to measure pressure-flow relationships over a wide
range to demonstrate a change in the slope (dP/dQ) to be
certain that a given drug or stimulus has caused an active
vasomotor change. As a matter of fact, to any point of the
pressure-flow diagram corresponds a given value of their ratio,
i.e. of the pulmonary vascular resistance. On the other hand,
a change in the slope of the pressure-flow relationships may be
due to recruitment or derecruitment of vessels and, in turn,
this phenomenon may be not just a passive gravity-dependent but
also an active physiological process. Hence, the actual
problem is not whether to use the pressure-flow diagram or to
compute the pressure-flow ratio, but rather to consider these
parameters together with all the possible variables known to
influence them. Doing so, we may realise that sometimes it is
not impossible, even in man, to produce circumstantial evidence
as to the nature of a change in pulmonary vascular resistance.

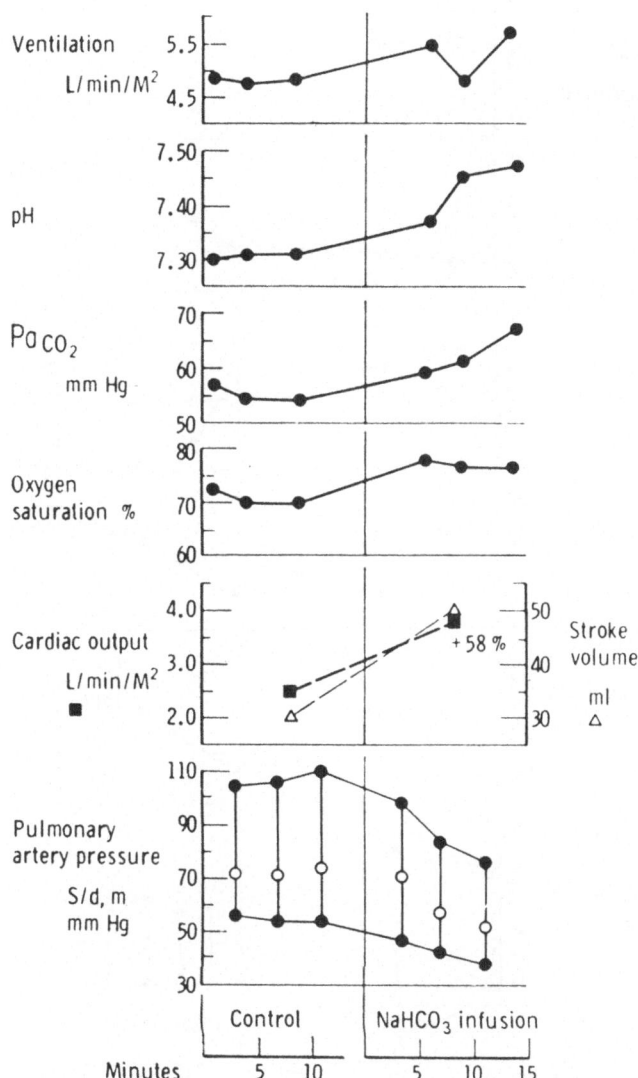

FIGURE 3:   Graphic representation of the effects of sodium
              bicarbonate in a patient with chronic bronchitis
              and cor pulmonale. From reference (24).

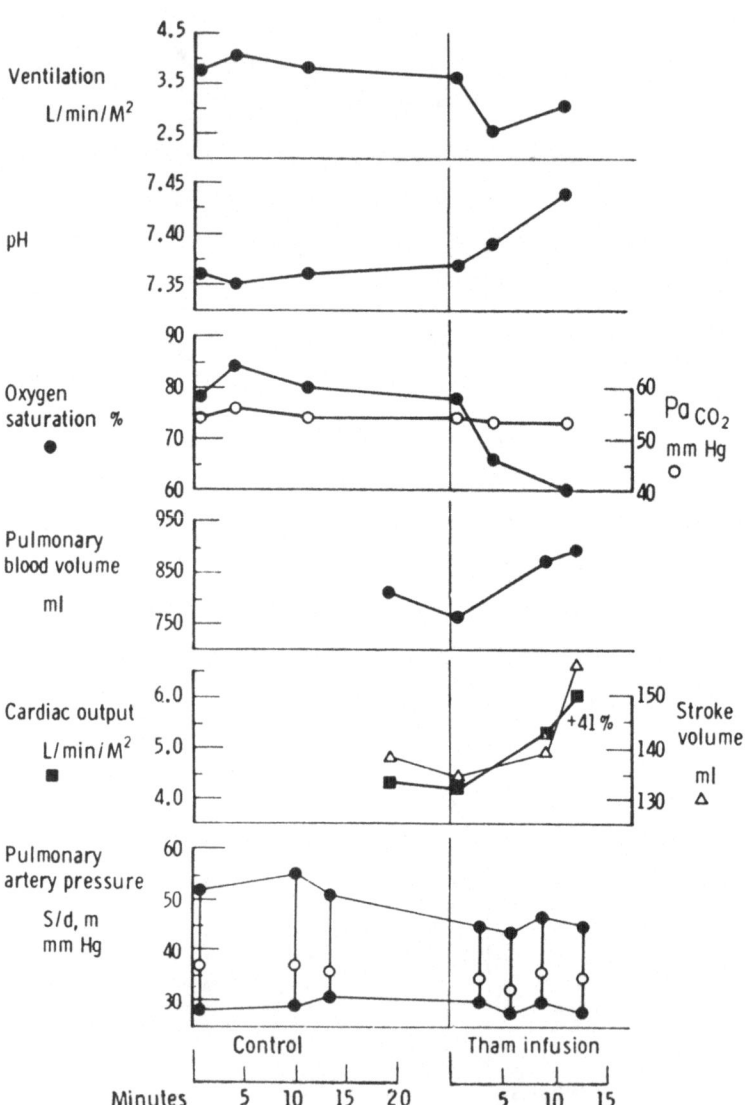

FIGURE 4:   Graphic representation of the effects of Tris
             (THAM) in a patient with chronic bronchitis and
             cor pulmonale.   From reference (24).

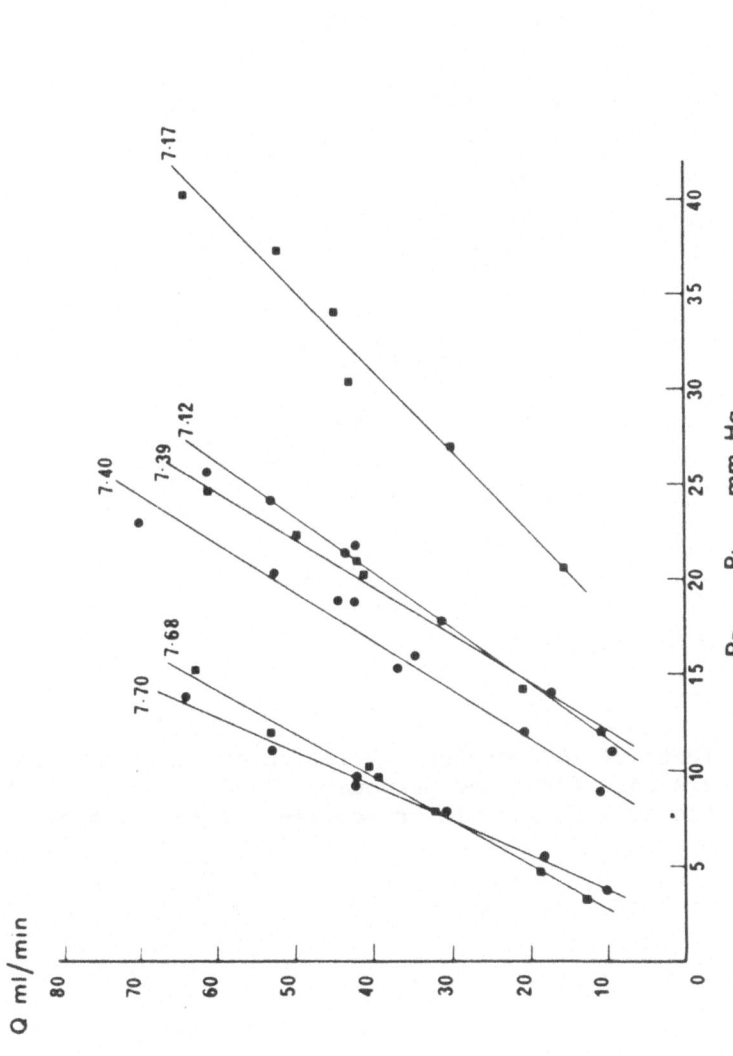

FIGURE 5:   The effect of hypoxia and changes in blood pH on pulmonary vascular resistance in an autoperfused left lower lobe of lung in vivo in a cat.   Abscissa, inflow pressure (P_{Pa}) to lobe minus outflow pressure (P_{La}); ordinate, blood flow to lobe.   Circles: ventilation with oxygen.   Squares: ventilation with 10% oxygen.   Blood pH, altered by infusion of acid and alkali, is shown.   From reference (1).

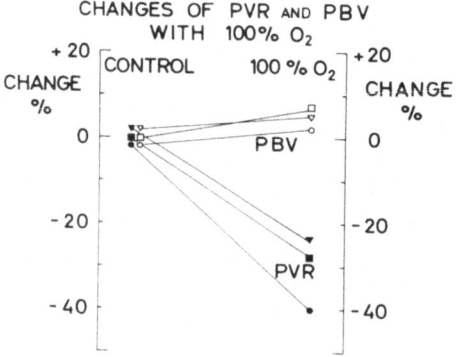

FIGURE 6:   Effects of 100% oxygen inhalation on pulmonary vascular
             resistance (PVR) and pulmonary blood volume (PBV) in 3
             patients with chronic bronchitis and cor pulmonale.

In man under physiologic conditions, and even more in many pathologic states, recruitment is not so important as in the experimental animal and the isolated lung. One example of this is offered by the following study (Figure 3). In a patient with pulmonary arterial hypertension secondary to chronic bronchitis and respiratory insufficiency, sodium bicarbonate infusion induced decrease of an elevated pulmonary artery pressure and rise of pulmonary blood flow (24). In this situation the change, i.e. decrease, of the pressure-flow ratio was interpreted as caused by an active vasomotor change and was ascribed to elevation of oxygen tension and decrease of hydrogen ion concentration. A similar study, where the buffer THAM was infused (Figure 4), suggested that decrease of the pressure-flow ratio could be ascribed just to lowering of the hydrogen ion concentration since, in this instance, oxygen tension diminishes (24). These conclusions reached by such studies in man, have been confirmed by a large variety of experiments in many animal species (Figure 5). This kind of regulation seems to influence to some extent also the capacity vessels of the lung - slight increment of pulmonary blood volume on breathing 100% oxygen but the effect on pulmonary vascular resistance is much greater (Figure 6). This response pattern, from the quantitative point of view, is opposite to that observed in presence of sympathetic stimulation.

4. Conclusions

We may then conclude that, whereas sympathetic stimulation influences pulmonary vascular compliance more than resistance, changes of oxygen tension and hydrogen ion concentration affect pulmonary vascular resistance more than compliance even though both characteristics of the pulmonary circulation are modified.

Finally, it should be added that, whereas in the presence of sympathetic stimulation there may be the problem of imposing an increased load to the right ventricle to prevent development of pulmonary oedema, in conditions of hypoxia and acidosis there is the problem of the need to divert blood flow from the pulmonary regions where hypoxia and acidosis is most marked.

REFERENCES

1. Barer, G. R. (1981).
   The physiology of the pulmonary circulstion and methods
   of study.
   In: Respiratory Pharmocology, ed. J. G. Widdicombe,
   Pergamon Press, Oxford, p.345.

2. Weibel, E. R. (1963).
   Morphometry of the human lung. Springer, Heidelberg.

3. Sobin, S., Tremer, M. H., Fung, Y. C. (1970).
   Morphometric basis of the sheet flow concept of the
   pulmonary alveolar microcirculation in the cat.
   Circ. Res., 26, 397-414.

4. Richardson, J. B. (1979).
   Nerve supply to the lungs. State of the Art.
   Am. Rev. Respir. Dis., 119, 785-802.

5. Spencer, H., Leof, D. (1964).
   The innervation of the human lung.
   J. Anat. 98, 599-609.

6. Hung, K. S., Mertweck, M. S., Hardy, J. D., Looslie, C. G.
   (1972)
   Innervation of pulmonary alveoli of the mouse lung: an
   electron microscopic study.
   Am. J. Anat., 135, 477-496.

7. Muller, L. R. (1911).
   Beitrage zur Anatomie, Histologie und Physiologie des
   Nervus Vagus, zugleich ein Beitrag zur Neurologie des
   Herzens, der Bronchien und des Magens.
   Dtsch. Arch. Klin. Med., 101, 421.

8. Verity, M. A., Bevan, J. A. (1968).
   Fine structural study of the terminal effector plexus,
   neuromuscular and intermuscular relationships in the
   pulmonary artery.
   J. Anat., 103, 49-63.

9. Hebb, C. (1969).
   Motor innervation of the pulmonary blood vessels of
   mammals.
   In: The Pulmonary Circulation and Interstitial Space.,
   ed: A. P. Fishman, H. H. Hecht, Univ. Chicago Press,
   p.195.

10. Bergofsky, E. H., (1980).
    Humoral control of the pulmonary circulation.
    Ann. Rev. Physiol., 42, 221-233.

11. Guintini, C., Maseri, M., Bianchi, R. (1966).
    Pulmonary vascular distensibility and lung compliance
    as modified by dextran infusion and subsequent atropine
    injection in normal subjects.
    J. Clin. Invest., 45, 1770-89.

12. Ingram, R. H., Szidon, J. P., Skalak, R., Fishman, A. P.
    (1968).
    Effects of sympathetic nerve stimulation of the
    pulmonary arterial tree of the isolated lobe perfused
    in situ.
    Circ. Res., 22, 801-15.

13. Pace, J B., Cox, R. H., Alvarez-Vara, F., Karreman, G. (1972).
    Influence of sympathetic nerve stimulation on puomonary
    hydraulic input power.
    Am. J. Physiol., 222, 196-201.

14. Gorton, R., Gunnells, J. C., Weissler, A. M., Stead, E. A.Jr.
    (1961).
    Effects of atropine and isoproterenol on cardiac
    output, central venous pressure, and mean transit time
    of indicators placed at three different sites in the
    venous system.
    Circulat. Res., 9, 979.

15. Aviado, D. M., (1965).
    The Lung Circulation.
    Oxford, Pergamon Press. vol. 1, pp. 343-355.

16. Gorlin, R., McMillian, I.K.R., Medd, W. E., Metthews, M. B.,
    Daley, R. (1955).
    Dynamics of the circulation in aortic valvular disease.
    Am. J. Med., 18, 855.

17. Segel, N., Harris, P., Bishop, J. M. (1960).
    The effects of synthetic hypertensin on the systemic
    and pulmonary circulations in man.
    Clin. Sci., 20, 49.

18. Morkin, E., Levine, O. R., Fishman, A. P. (1964).
    Pulmonary capillary flow pulse and the site of
    pulmonary·vasoconstriction in the dog.
    Circ. Res., 15, 146.

19. Marazzi, A. (1939).
     Electrical studies on the pharmacology of autonomic
     synapses. I. The action of parasympathomimetic drugs on
     sympathetic ganglia.
     J. Pharmacol. Exp. Ther., 65, 18.

20. Cahen, R. L., Tvede, K. M. (1953).
     Action of atropine on sympathetic ganglia.
     Arch. Int. Pharmacodyn, 94, 248.

21. Giuntini, C., Maseri, A., Mariani, C., Contini, C., Donato, L.
     Right and left ventricular performances as modified by
     dextran infusion and subsequent atropine injection in
     normal subjects.
     In Preparation.

22. DeGeest, H., Levy, M. N., Zieske, H., Lipman, R. I. (1965).
     Depression of ventricular contractility by stimulation
     of the vagus nerves.
     Circ. Res., 17, 222.

23. Levy, M. N., Lipman, M., Ng, R. I., Zieske, H. (1966).
     Vagus nerves and baroreceptor control of ventricular
     performance.
     Circ. Res., 18, 101.

24. Enson, Y., Guintini, C., Lewis, M. L., Morris, T. Q.,
     Ferrer, M. I., Harvey, R. M. (1964).
     The influence of hydrogen ion concentration and hypoxia
     on the pulmonary circulation.
     J. Clin. Invest, 43, 1146-1162.

DISCUSSION

SPEAKER: GUINTINI   CHAIRMAN: CUMMING

DENISON: Carlo, I want to ask you in which posture your experiments with atropine were done, the reason being to follow up the fundamental observation that when you stand up some lung vessels are empty and when you lie down you can see radiographically that the basal vessels do not increase in size. The pulmonary bed does collapse at the apex when erect but is distended during recumbancy so that the compliance of the pulmonary circulation over that very small pressure range is quite large but it doesn't stretch very much, so it matters quite a lot whether you were lying down or sitting up when you did your experiments.

GUINTINI: The studies were done in the supine position both the control, the dextran infusion, and the post-atropine determinations.

DENISON: That means they were done when the circulation was already full.

GUINTINI: That's right.

DENISON: So when you are speaking of the compliance you are speaking of the compliance of the distended pulmonary circulation.

GUINTINI: I would agree with you to this extent that when we have measured pulmonary blood volume in the sitting position, we found it to be 30% less than in the supine position. As soon as bicycle exercise begins, in the sitting position the blood immediately increases to the lying value and continuing exercise scarcely increases it further. Apparently in this situation some other factor is operative, since the pulmonary blood volume obtained following Dextran infusion were 50% greater than those in the supine position.

CHAIRMAN: What are your current views on the distensibility of the pulmonary circulation?

GUINTINI: My present working hypothesis is as follows:- when
                 the volume or pressure of blood in the pulmonary
                 circulation is increased, mechanisms are invoked
                 which are quite different to those associated with
                 an increase in flow. In the first case there is a
                 limitation in the hydraulic power of the right
                 ventricle and in the latter it is enhanced and I
                 think there exist specific sensory receptors for
                 thes two situations. Thus if we study the
                 circulation in the first mode we conclude that it
                 is distensible and it is in this mode that the
                 majority of studies have been carried out.
                 Studies in the second mode would come to the
                 conclusion that the pulmonary vascular bed is not
                 distensible.

LEE: How do you see the connection between the data of
                 Ron Linden on vagal and sympathetic efferents from
                 the pulmonary veins, and their effects on the
                 systemic circulation since these two parts of the
                 circulation must work together. Linden points out
                 that efferent activity produced by distension of
                 the pulmonary veins releases anti-diuretic hormone
                 and changes vascular tone in the systemic
                 circulation and both will affect circulating blood
                 volume drastically.

GUINTINI: This is a difficult subject, but I believe that
                 the relationship between such efferents and
                 circulating blood volume has not been convincingly
                 established. Anatomical studies certainly confirm
                 the pressure of receptors in large pulmonary veins
                 so they presumably have a function. Short term
                 effects have been demonstrated by, so far as I am
                 aware, the longer-term effects leading to changes
                 in blood volume have not been shown.

ASSESSING THE EFFECTS OF DRUGS ON VENTILATION/PERFUSION

D. M. Denison, J. F. Waller, E. Sawicka, J. Bowyr & C. Busst

The Lung Function Unit

Brompton Hospital, London SW3 6HP

The lung is usually thought of as a passive window for metabolic gas exchange but, of course, this is not exactly so, as there is an active and constantly adjusted matching of ventilation to perfusion and vice-versa. Sometimes the processes become hyperactive, for example in asthma or pulmonary hypertension, and at other times they become weakened by disease as in emphysema or multiple micro-embolization. They can also be modified by treatment. However, it is by no means easy to measure or express these effects, when they influence more than one aspect of lung function at the same time.

We have found it helpful to tackle this problem pictorially and would like to begin this communication with some illustrations of that approach. Table 1 compares some observed and predicted attributes of the lungs of a 26 year old North Sea Diver who had ruptured his lungs on two occasions during emergency ascents from depth. The numbers given are fairly uniformative to most clinicians and are not especially easy to assimilate or interpret when presented in this way. Figure 1 shows the same set of results in pictorial form. It depicts the observed maximal flow-volume loop scaled in terms of predicted Total Lung Capacity (TLC) and predicted Forced Vital Capacity per second (FVC/sec) on the respired volume and respired flow axes respectively. The solid dot marks the actual Forced Expired Volume in one secong (FEV1). The origin of the loop is set at the observed TLC, so the loop can be compared with that predicted for a healthy lung in a normal person of the same age, sex and height. This is shown by the broken lines. It is immediately apparent that this man has

TABLE 1:   The results of the routine function tests on the
           lungs of the 26 year old diver, that are summarised
           in Figure 1.

FEV1	ml	3470 – 4690	4640
FVC	ml	4140 – 5600	5460
FEV1/FVC%		76 – 88	85
TLC	ml	5330 – 7210	6370
RV	ml	1430 – 1940	1170
FRC	ml		3140
VA	ml	4950 – 6710	5820
VC	ml	4140 – 5600	5200
IN.AWR	SI	<0.2	< 0.2
SGAW	SI	1.3 – 3.6	>3.6
DLCO	SI	9.63 – 13.03	9.14
KCO	SI	1.58 – 2.14	1.57

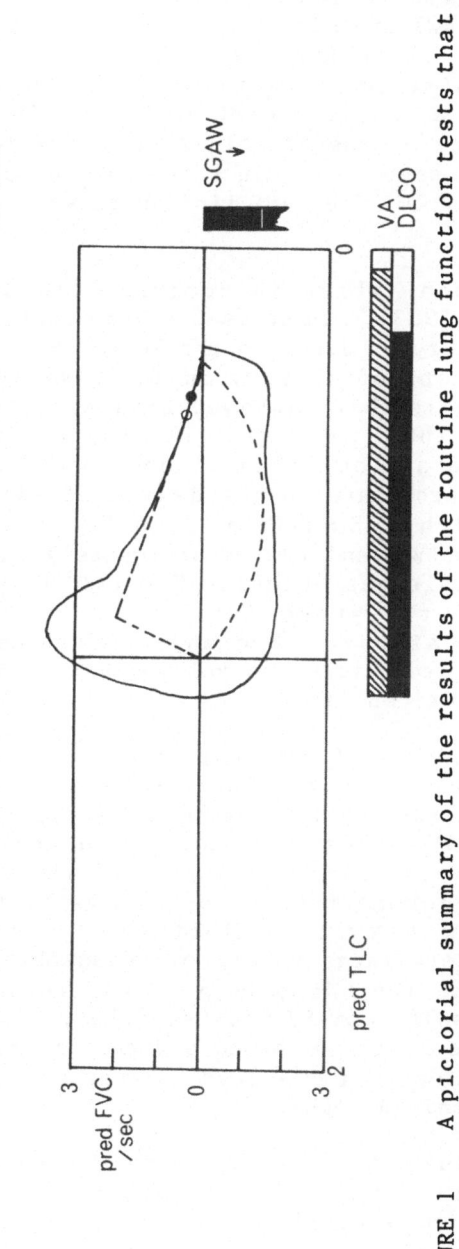

FIGURE 1   A pictorial summary of the results of the routine lung function tests that are listed in Table 1.

super-normal spirometry, a slightly higher TLC and lower RV than predicted, while his FVC and FEV1/FVC ratio are substantially better than predicted. The shaded bar to the right of the main rectangle indicates that specific airway conductance (SGAW) exceeds the lower limit of normal (shown by the transverse white bar), and the shaded parts of the two horizontal bars below the main rectangle show the fraction of TLC reached by a single breath of Helium in 10 seconds (VA), and the quantity of haemoglobin that was accessible to carbon monoxide in the same time. In this case the breath invades the lung very easily but does not find quite as much haemoglobin as it should.

This approach, which is described in detail elsewhere (Denison et al 1983), can be used to summarise the functional effects of surgery or drugs on the lung. For example, Figure 2 shows the effects of surgery on an 11 year old girl with a tracheal stenosis that followed intubation for pre-existing lung disease. The upper panel illustrates her pre-operative state. She had a flow-volume loop typical of rigid airways obstruction, appropriate to the degree of airway conductance limitation, with a low TLC and high RV, poor invasion of the lung in ten seconds and disproportionately little haemoglobin in the accessible gas volume. The bottom panel shows that dilatation of the stenosis, as expected, improved the spirometry and airway conductance but had no effect on total lung capacity, accessible gas volume (VA) or whole lung carbon monoxide transfer (DLCO).

Figure 3 shows the functional effects of two inhalations of salbutamol aerosol on the lungs of a 43 year old man with asthma. Before the inhalations he had a clearly raised TLC and RV and a moderate pressure-dependent collapse of the airways, as evidenced by the expiratory limb of his flow-volume loop. He maintained normal inspiratory flows by increased effort despite a low airway conductance, and as a result a disproportionately large volume of accessible haemoglobin was sucked into the lung despite good gas invasion. After the salbutamol, his TLC and RV fell somewhat, airway collapse was somewhat less evident on forced expiration, airway conductance and gas space invasion improved but the volume of accessible haemoglobin remained high.

Figure 4 illustrates a rather more obvious improvement in a 47 year old man with severe asthma who was exposed to bronchodilator and steroid therapy for 18 months.

He shows considerable falls in TLC and RV associated with very significant increases in FEV1 FVC, airway conductance,

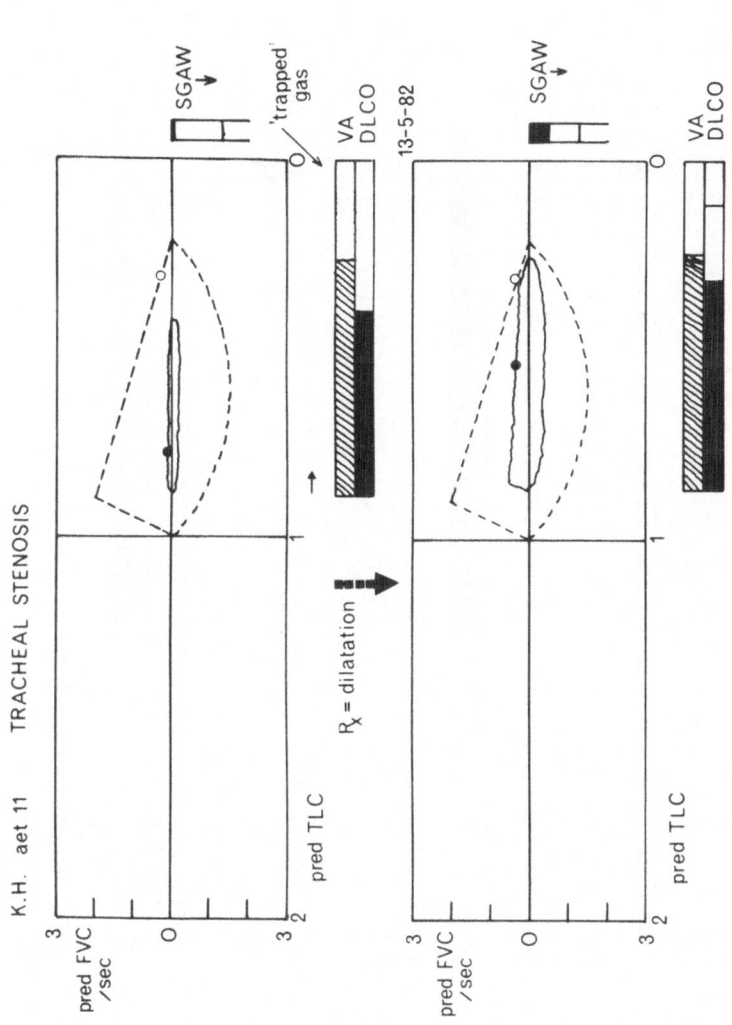

FIGURE 2   The pre and post-operative findings on an eleven year old girl who had her tracheal stenosis dilated.

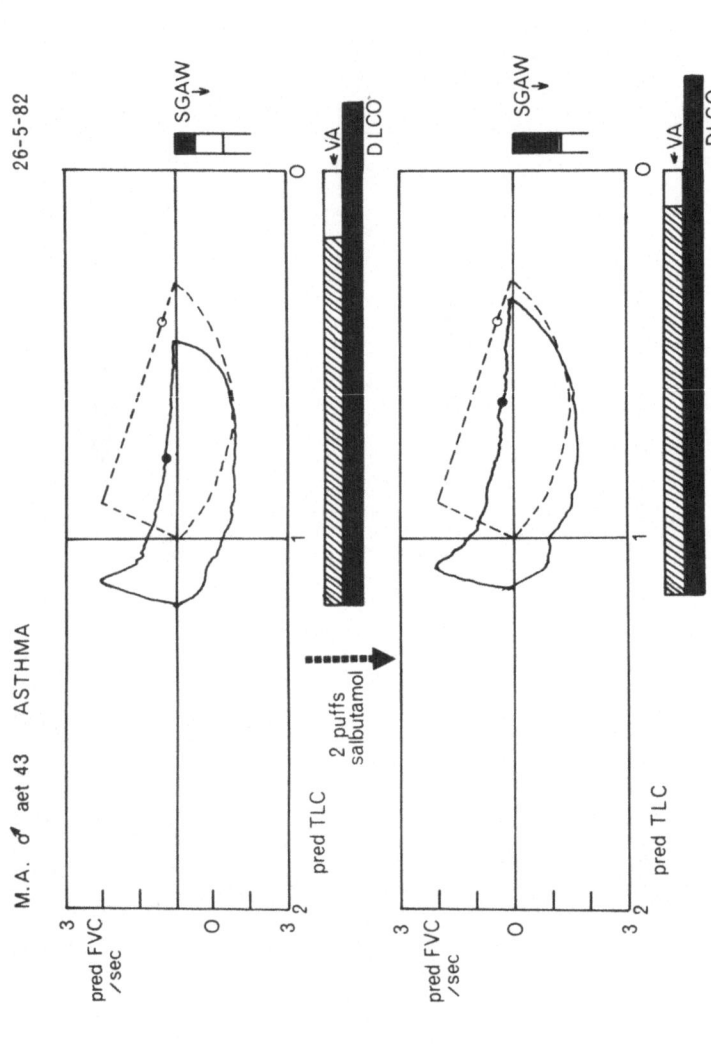

FIGURE 3   The functional effects of two puffs of salbutamol on the lungs of a 43 year old man with asthma.

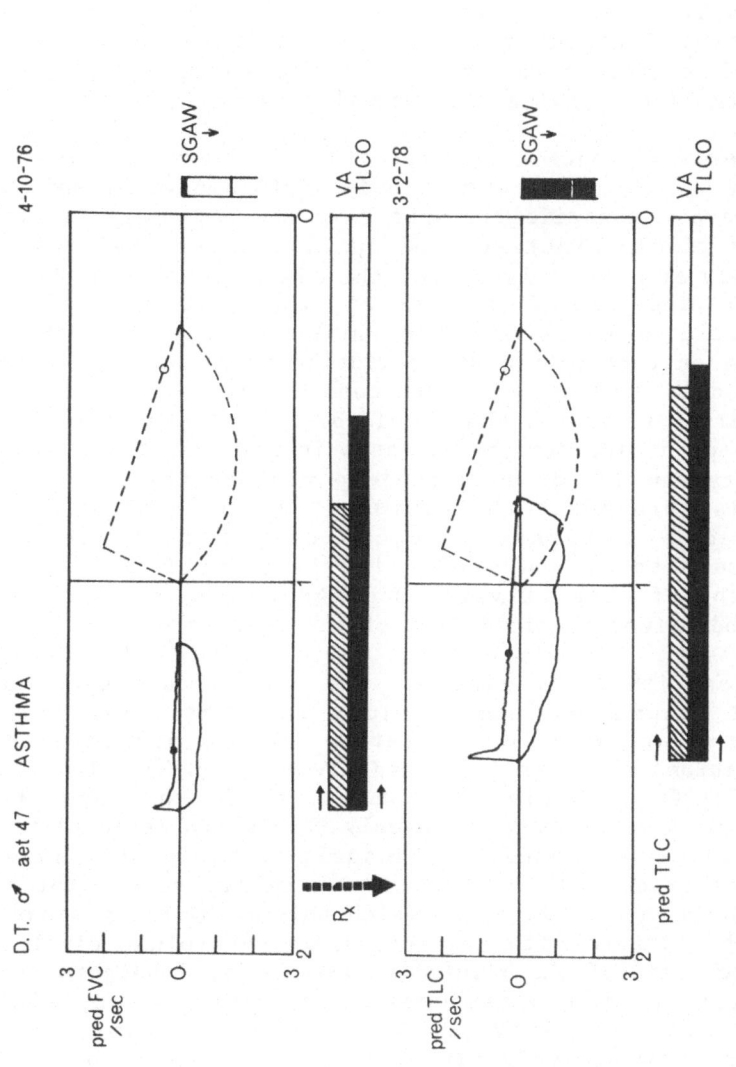

FIGURE 4   The functional effects of prolonged broncho-dilator and steroid therapy on the lungs of a 47 year old man with asthma.

accessible gas volume and accessible haemoglobin.

Figure 5 show the effects of twelve weeks steroid therapy in a 25 year old lady with extrinsic allergic alveolitis. Initially her lungs were small and short of accessible haemoglobin. After treatment her lungs re-expanded to their normal size, without much limitation to gas invasion, the amount of accessible haemoglobin increased dramatically and the flow volume loop regained its normal form.

The tests of routine lung function that have been described, i.e. flow volume loops, whole body plethysmography and single breath carbon monoxide transfer have the advantages that they are very widely practised and easily interpreted against a large body of normal data, but the disadvantage is that they take something like 3/4 hour to complete. It would obviously be desirable to have some other simpler tests of the effects of treatment on ventilation and perfusion which was reliable and that could be repeated at short enough intervals to follow the time course of events, that could be interpreted easily. The measurement of arterial blood gases is one such test. However, it is often unwise or unethical to measure them at frequent intervals throughout the course of a drug trial. We have developed an alternative technique, the single breath argon/freon test, and would like to illustrate its use to assess the effects of various drugs on the accessible gas volume and effective blood flow of the lung in man.

The test itself is simple. The patient breathes out to residual volume and then rapidly inspires a single vital capacity of an oxygen-rich mixture that is marked with low concentrations of argon (10%) and freon 22 (3.5%). The patient breatholds for a couple of seconds at total lung capacity and then breathes out slowly and evenly through to residual volume. Expired air is analysed continuously at the mouth, normally providing traces of the form shown in Figure 6. The argon trace (a) shows the extent to which this insoluble gas has been diluted by mixture with the accessible gas volume of the lung (VA). The ratio of the diluted to the inspired heights of this concentration are proportional to the ratio of accessible to inspired volume (VA/VI). The freon concentration (f) diminishes progressively with inspiration because this soluble gas is removed by the effective pulmonary blood flow (Q) and the ratio of blood flow to the accessible gas volume (Q/VA) is easily calculated by comparison of the two tests. This technique has been described in detail elsewhere (Denison et al 1980). The values that we find in a group of normal subjects are given in Table 2, which is taken from Waller (1982).

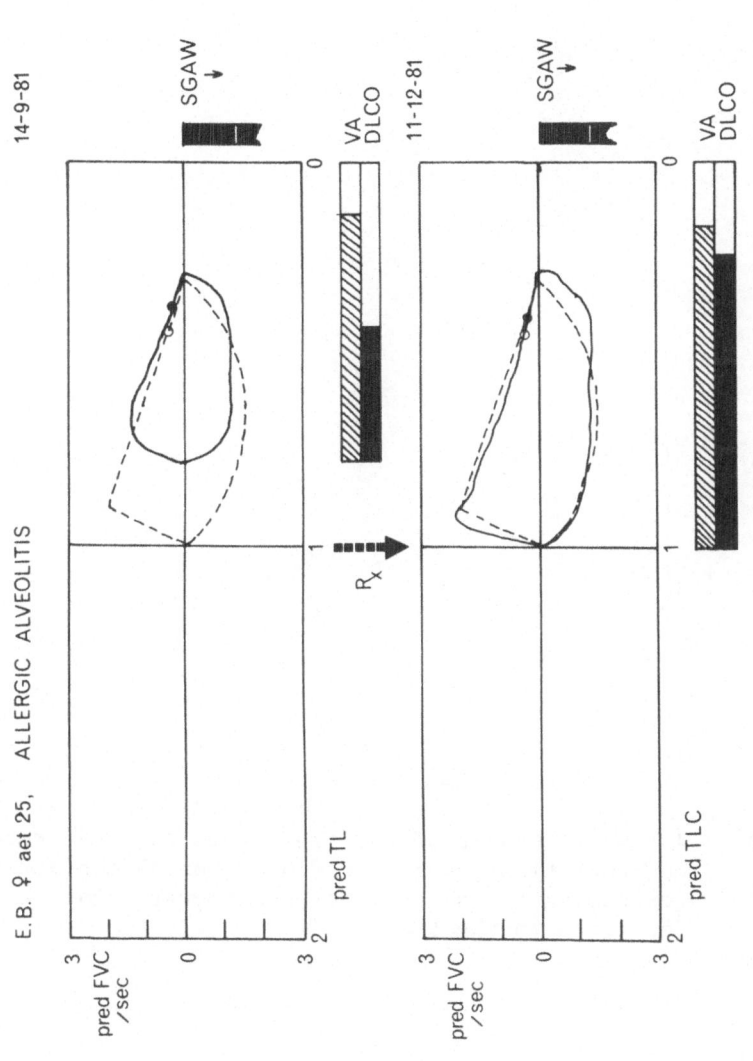

FIGURE 5   The functional effects of steroid therapy on the lungs of a 25 year old lady with allergic alveolitis.

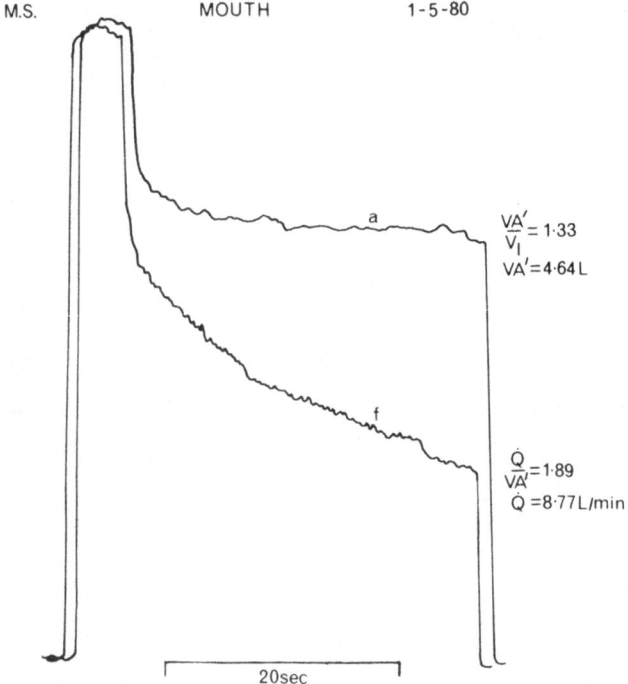

FIGURE 6   A typical tracer of respired argon (a) and freon
concentrations, obtained from a normal subject
performing the single-breath manoeuvre that is
described in the test.

TABLE 2.   The results of the single-breath argon-freon test
            in 44 normal subjects (28 male, 16 female);   mean
            age: 35 years (range 23 - 66).

Accessible residual volume* 1.58 ± 0.41 L

Accessible lung water 0.28 ± 0.13 L

Effective pulmonary blood flow 5.44 ± 1.05 L

NB: *   Accessible gas volume (VA) = vital capacity plus
          accessible residual volume.

In diseased lungs the treatment of argon and freon can be
very different, as can be seen in two cases shown in Figure 7.
The distinction between normal and abnormal people are often
quite striking as is shown by the rebreathing variant of the
test which is suitable for use in children, (Figure 8).   The
studies of Waller (1982) in particular, show that this test can
be repeated at intervals as short as 3 to 4 minutes, obtaining
values that are repeatable within + - 5%.   This has enabled us
to demonstrate the time course of the effects of various drugs
on the lung.   These are illustrated in Figures 9 - 12.

The first of these concerns the relative effects of oral and
inhaled terbutaline on the lung.   In a double-blind crossover
trial, eight patients with moderately severe stable asthma were
given incremental doses of terbutaline by one and then the
other route while VA + Q were determined, in dose-response
fashion.   The terbutaline caused both to rise (Figure 9) to an
extent independent of the route (Pierce et al 1981).

In a second study eight men and four women with stable long-
standing asthma and an FEV1 that rose by 25% or more on the
inhalation of salbutamol, were exposed to 2 inhalations of 2.5
mgs of Terbutaline and a placebo tablet, or a tablet of 2.5 mgs
Terbutaline plus two placebo inhalations;   in a double-blind
cross-over trial.   Measurements of VA and Q were taken on five
occasions before medication and in quintriplicate again at 60,
90 and 120 minutes after the dose.   After the 120 minute
observations the patients received two inhalations of 0.25 mgs
Terbutaline and the five recordings were repeated again 30
minutes later.   The results of this study are summarised in
Figure 10.   In essence it shows that oral and inhaled
terbutaline have similar effects on the accessible gas volume,
but in these almost equivalent doses, administration by the
oral route produces significantly greater increases in
effective pulmonary blood flow.

Another study was designed to investigate whether two agents
(Prizidolol and Propranalol) produced any significant changes
in airway function or effective pulmonary blood flow in normal
subjects.   With the doses used, neither agent induced any
changes in those variables thought to reflect central or
peripheral airway function, but the Prizidolol produced a
significant increase in effective pulmonary blood flow, as
shown in Figure 11.

In a fourth study, eight normal subjects were given a single
intravenous dose of 40 mg Prednisolone, and an equal volume of
intravenous saline as placebo, in a double-blind trial.   The
prednisolone produced significant increases in effective

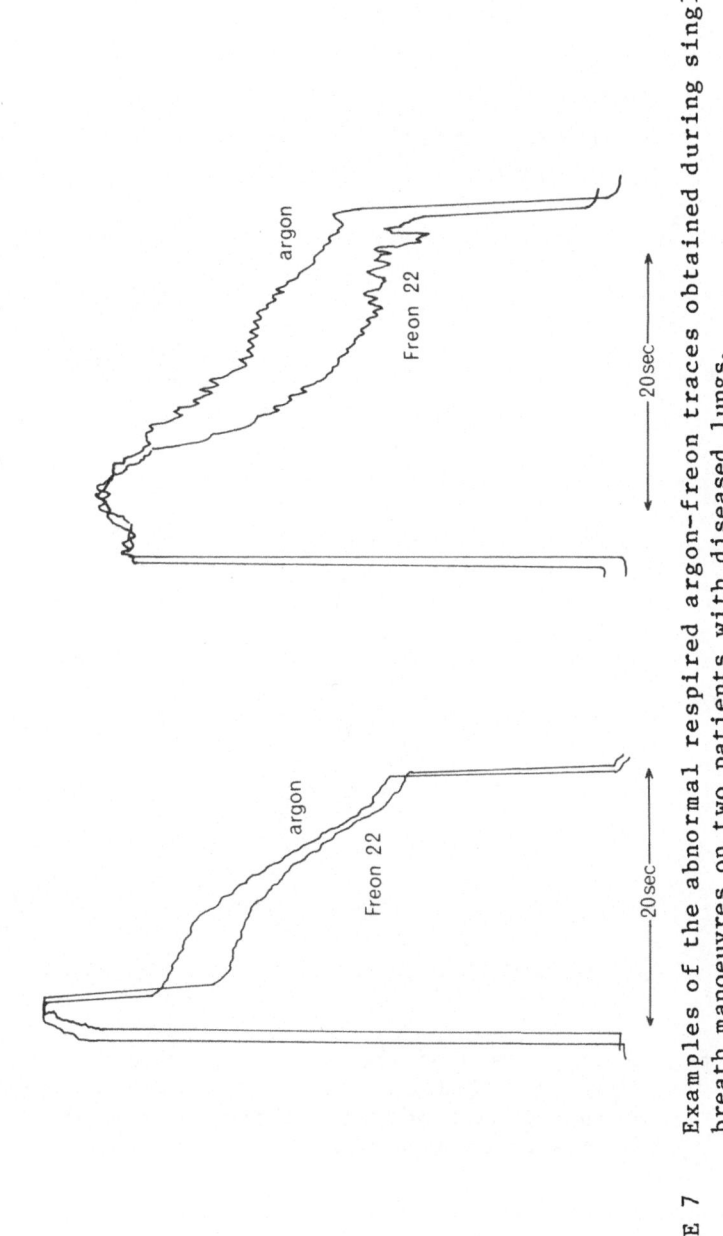

FIGURE 7   Examples of the abnormal respired argon-freon traces obtained during single-breath manoeuvres on two patients with diseased lungs.

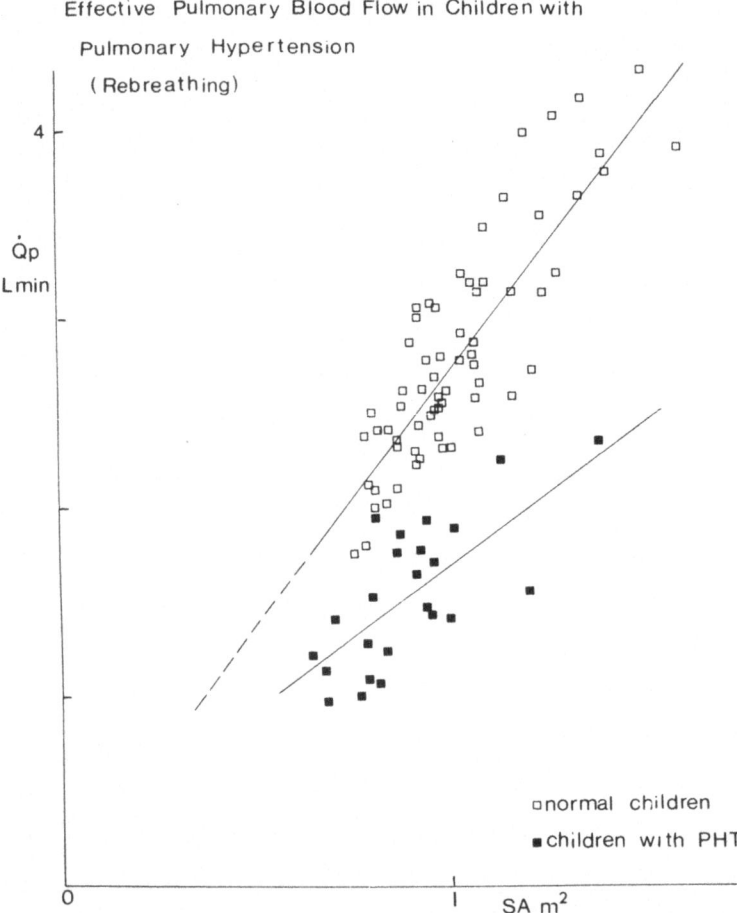

FIGURE 8   A comparison of the effective pulmonary blood-flow
             (Qp) in children with pulmonary hypertension
             compared with normal children of comparable whole-
             body surface area (SA).

A comparison of the effects of intravenous and inhaled terbutaline on VA and Qp

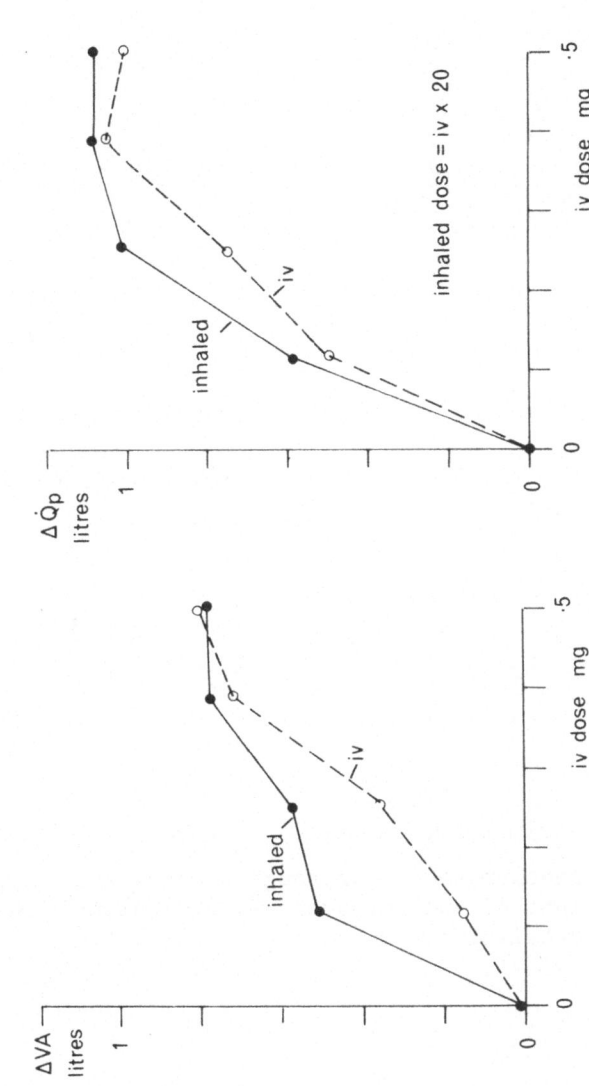

FIGURE 9 The effects of incremental doses of inhaled and intravenous terbutaline on accessible volume (VA) and effective pulmonary blood flow (Qp) in eight patients with asthma. Note the different doses employed in the two routes.

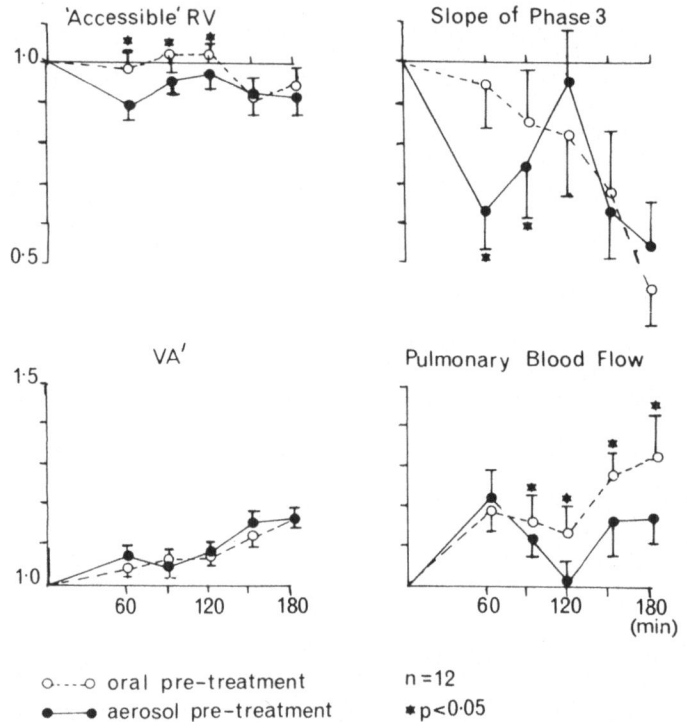

FIGURE 10   Another study, on roughly comparable single-doses of
             inhaled and intravenous terbutaline. See text for
             details.

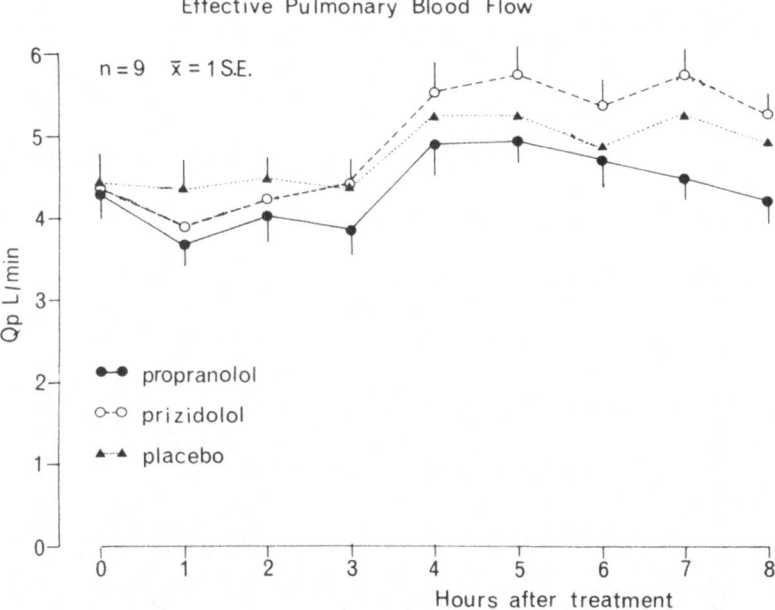

FIGURE 11   A comparison of the effects of oral propranolol, oral prizidolol and an oral placebo on effective pulmonary blood flow.   The placebo trace reveals the normal
circadian variation in this measure.

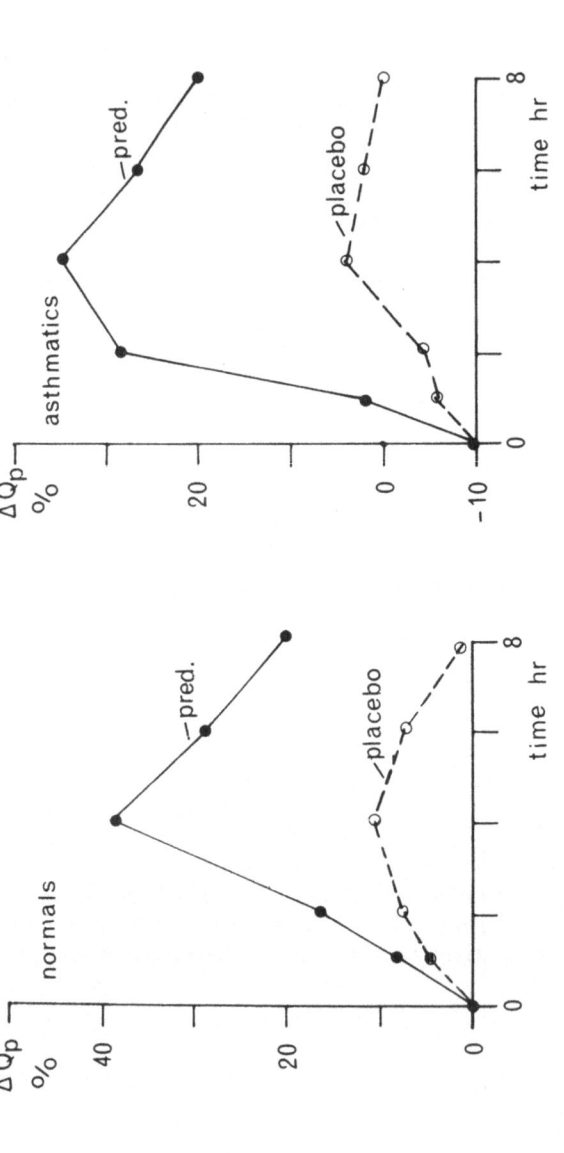

Effects of 40 mg intravenous prednisolone on Q̇p in normals & asthmatics

FIGURE 12   The effects of a single-dose of prednisolone on effective pulmonary blood flow in eight healthy adults.

pulmonary blood flow, but not accessible gas volume in these subjects, whereas 10 stable asthmatics showed increases in both, (Fig. 12).

These studies suggest that the single-breath argon/freon test is capable of following the time-courses of drug-induced changes in alveolar gas mixing and effective pulmonary blood flow and is therefore one suitable method of observing ventilation/perfusion adjustments during drug trials on the lung.

REFERENCES

BOYER, J.
        MD. Thesis: The non-invasive measurement of pulmonary blood flow in children.
        University of Oxford, 1983.

DENISON, D. M., Davies, N. J. H., Meyer, M., Pierce, R. J. and Scheid, P. (1980).
        Single exhalation method for study of lobar and segmental lung function by mass spectrometry in man.
        Respir, Physiol., 42, 87-99.

DENISON, D. M., DuBois, R., and Sawicka, E. (1983).
        Does the lung work
        Brit. J. Dis. Chest., 77, 35-50.

EDMONSTONE, M. R., Waller, J. F., Manghani, K. K., Hill, P. and Bell, A.
        Pulmonary effects in man of oral prozidolol hydrochloride (SK & F 92657): a new antihypertensive agent.
        (Submitted to Brit. J. Clin. Pharmacol.)

PIERCE, R. J., Payne, C. R., Williams, S. J., Denison, D. M., and Clarke, T. J. H. (1981).
        Comparison of intravenous and inhaled terbutaline in the treatment of asthma.
        Chest, 79, 506-511.

SHAW, R. J. S., Waller, J. F., Hetzel, M. R. and Clark, T. J. H. (1982).
        Do oral and inhaled terbutaline have different effects on the lung.
        Brit. J. Dis. Chest., 76, 171-176.

WALLER, J. F. (1982).
        Assessment of a single-breath test of whole and
        regional lung function in man.
        University of London MD Thesis.

THE ASSESSMENT OF DRUGS ACTING ON

INTERSTITIAL LUNG DISEASES

R. K. Knight

Brompton Hospital

Frimley, Camberley, Surrey

There are many causes of interstitial lung disease. I will consider the methods that have been developed and evaluated to assess and follow patients with cryptogenic fibrosing alveolitis with some reference to sarcoidosis. The techniques used in these conditions can be widely applied to other conditions as appropriate.

Cryptogenic fibrosing alveolitis (CFA) is diagnosed by excluding:-

1. patients in whom an extrinsic cause can be found.

2. those with precipitins to Micropolyspora faeni or avian antigen.

if (a) a biopsy is available, the features are:-
fibrosis of alveolar walls and a variable interstitial/intra-alveolar infiltrate with no granulomata or features of pneumoconiosis.

or (b) a biopsy is not available, the following are essential diagnostic features:-

(i)   widespread persistent radiographic shadowing

(ii)   widespread crackles on auscultation

Supportive criteria are finger clubbing and a restrictive ventilatory defect.

239

Many workers in America use the term idiopathic pulmonary fibrosis for the same condition. Others still separate desquamative interstitial pneumonia (DIP), where the biopsy shows a relative uniformity of lesions throughout the tissue, a sparse interstitial cellular infiltrate, the prominent lining of alveoli by large round cells and abundant filling of small air spaces by mononuclear cells, from usual interstitial pneumonia (UIP). In the latter, more common condition, the biopsy shows a highly varied structure from normal alveolar walls to fibrotic, end-stage lesions; a dense pleomorphic interstitial infiltrate including many lymphocytes and monocytes with few cells in air spaces. In Britain these conditions are considered to be part of the spectrum of CFA with a cellular biopsy being found in about 19% of cases.

There is good morphological evidence that the initial lesion in all forms of pulmonary fibrosis is an alveolitis. This inflammatory stage may or may not interfere with the structural function of the lung sufficiently to cause symptoms by itself. It may be short-lived and be followed by the development of pulmonary fibrosis or may co-exist with the fibrotic process. The end result may be progressive impairment of ventilatory function often leading to premature death.

The large functional reserve of the lung means that when symptoms occur the disease is advanced. The main drugs used in this condition (steroids and immunosuppressants) have dangerous side-effects. As yet we can do little, if anything, to modify fibrous tissue once it has formed. We therefore need to know the activity of the alveolitis on presentation and throughout treatment so that it can be suppressed optimally by the smallest dose of medication.

METHODS OF EVALUATION

Lung Biopsy

This is the absolute way, in life, of confirming the diagnosis and assessing the relative activity of the alveolitis and fibrotic process.

Method

Open lung biopsy under general anaesthetic (1)

A small (8 - 10 cm) submammary incision is used and a wedge of lung tissue removed from both a macroscopically abnormal area and from an area minimally involved. The specimen is divided and processed for light, electron microscopy and, occasionally, immunofluorescent studies. Out of 1174 open lung biopsies in the world literature (2), a specific diagnosis was made in 82% of biopsies with complications occurring in 24.6% and an overall mortality of 0.68%.

Although this is the yardstick by which others are measured, many patients may present with severe ventilatory impairment which do not allow this major procedure. Other techniques have therefore been developed in the hope that they can be carried out under local anaesthetic with a lower morbidity.

Cutting needle biopsy

These needles are adaptations of the Silverman needle and are associated with a complication rate of over 50%. They should no longer be used.

Trephine lung biopsy

This technique (3) provides diagnostically useful material in up to 90% of cases (2). The commonest complication is pneumothorax (31%) but this rarely (16%) requires treatment. Recently in a study of 62 patients with CFA, the information provided was similar to that obtained by open lung biopsy (4).

Transbronchial lung biopsy

The development of this method by fibreoptic bronchoscopy is one of the major advances in pulmonary diagnosis. There are few complications (less than 5%), it can be carried out under local anaesthetic and may be repeated several times. In sarcoidosis, it is the lung biopsy technique of choice providing positive histology in 57% of patients with Stage I disease, 77% Stage II and 91% of Stage III disease (2). Unfortunately, it is not so helpful in CFA where the small sample size (2 x 2 mm) prevents the appraisal of the spectrum of disease that can be obtained from the larger sample provided by trephine or open biopsy.

In patients fulfilling the criteria of DIP the mean survival was 12.2 years and 61.5% improved with steroids (5). In the same series, patients with UIP had a mean survival of only 5.6 years and only 11.5% improved with steroids. Others, (4,6) have confirmed that patients with a more cellular biopsy and less fibrosis have a better prognosis and that these are the patients who respond to steroids.

Unfortunately, none of the lung biopsy techniques that have

been useful in assessing the degree of alveolitis and allowing some idea of prognosis, can be repeated and therefore used to monitor therapy. What other, less invasive, test can be used to assess the severity of the disease and monitor its subsequent progress with drug therapy?

Symptoms

Dyspnoea is the symptom we are trying to treat. Can we therefore grade it and use it to assess the severity of disease and its response to therapy?

The grade of dyspnoea at presentation is related to vital capacity (VC), severity of radiographic appearance and survival (7). This symptom, however, only occurs when the disease is advanced and does not allow separation of alveolitis from fibrosis.

Physical signs

In asbestosis, late inspiratory crackles are heard on auscultation before the chest x-ray becomes abnormal (8). They are therefore a sensitive indicator of disease. They become more widespread as the disease progresses but have not been used to assess its severity. It is unlikely they would separate alveolitis from fibrosis as they occur in 96% of all patients with CFA (7).

Chest x-ray

This can be analysed using the UICC criteria (9). The "profusion score" is the number of opacities/unit area of the radiograph. This correlates with a degree of dyspnoea at presentation and ultimate survival (7). Although CFA characteristically progresses from a "ground glass" appearance on the x-ray to a nodular, reticular or reticulo-nodular pattern and then on to a more coarse pattern with cystic areas and finally, a honeycomb pattern the severity of shadowing correlates poorly with morphological estimates of severity. Some patients with advanced disease on biopsy have a normal radiograph and even in honey-comb lung, there may be areas of active alveolitis amenable to therapy. The x-ray therefore gives only a rough idea about the disease and its effect on pulmonary architecture but as it is readily available and repeatable, it will form part of routine assessment and follow-up of drug therapy.

Pulmonary function

A variety of pulmonary function tests are used to assess and monitor treatment in interstitial lung disease. The tests that are routinely used include VC and/or total lung capacity and gas transfer. Resting arterial oxygen tension and exercise testing provide valuable additional information but are not used routinely in all cases. Most pulmonary function tests detect derangement of pulmonary structure and function and although useful in assessing and following patients, do not differentiate alveolitis from fibrosis.

Blood tests

Antinuclear antibody, rheumatoid factor, immunoglobulin levels and erythrocyte sedimentation rate may all be raised in CFA. None of these are, however, related to steroid responsiveness and to survival and probably do not reflect the intensity of alveolitis. The measurement of angiotensin converting anxyme (ACE) is used in sarcoidosis. This correlates with the number of sites of the body involved but does not correlate with the activity of the alveolitis as measured by more invasive tests.

Broncho-alveolar lavage (BAL)

Routine fibreoptic bronchoscopy is carried out after premedication with Papaveratum 10 - 20 mg and Atropine 0.6 mg. Supplemental O_2 is administered by a single pharangeal catheter as lavage drops arterial pO_2 by approximately 3kpA (22.7 mmHg). The segmental airways in the lower and middle lobes are assessed to see whether the end of the bronchoscope can be inserted 1 - 2 cms past the orifice and whether the bronchus collapses too easily on application of suction (if so too little fluid may be returned). A 60 ml aliquot of normal saline buffered to pH 7.0 (270 μL of 8.4% sodium bicarbonate to 500 ml of saline) is inserted into the chosen segment during inspiration. The patient is asked to take a deep breath after each aliquot and during expiration the fluid is aspirated into a siliconised glass trap bottle, in ice. The procedure is repeated until 100 - 150 ml of lavage fluid has been collected (10).

Contraindications

1. PaO_2 less than 70 mmHg or 9.3 kpA with supplemental O_2
2.   Myocardial infarction within the previous six months.
3.   Unstable angina
4.   Cardiac failure.

Complications

   These are minor and occur in less than 5% of cases. Transient fever, bronchospasm and nasal bleeding are the commonest. The procedure is well tolerated and can be repeated to follow the effect of drug therapy.

Results

   The technique samples the alveolar space and it has been calculated that 98% of the epithelial surface seen by the fluid is alveolar. Thus the contribution from the airways is negligible provided inflammatory airway disease does not coexist. The cell populations collected at lavage can be counted and the result expressed as a total number of each cell per ml of lavage fluid or as a differential cell count.

Total cell count

   This method shows that smokers have a three-fold cell yield compared with non-smokers. Unfortunately, it is impossible to get a uniform volume return — the cells are diluted in a variable volume in each patient. In view of this difficulty, differential cell counts are usually carried out to allow comparison with other patients and with the same patient on other occasions.

Differential cell count

   This is the commonest method used. Usually 200 – 500 cells are counted and a typical count for a normal person is:

Macrophages	93% ± 5%
Lymphocytes	7% ± 1%
Neutrophils	less than 1% (in cigarette smokers up to 5%)

The lymphocyte subpopulations in BAL fluid have a distribution similar to peripheral blood:-

   73% ± 4% are T cells of which 48% ± are Helper cells and 25% ± 5% Suppresor cells.

8% ± 3% are B cells of which less than 1% are spontaneously producing immunoglobulin.

In alveolitis, there is a change in cell population.

a)   There is an increase in the total number of inflammatory and immune effector cells.

b)   There is a shift of cell population to emphasise either neutrophils or lymphocytes.

c)   There is activation of one or more cell types, these activated cells may modulate granuloma formation, attract inflammatory cells and modulate the process of fibrosis.

Alveolitis with increased neutrophils

The cells consist of macrophages, lymphocytes and neutrophils. CFA and asbestosis are in this group. In CFA locally-produced immune complexes activate the macrophage via its IgGFc receptor and cause release of a chemotactic factor for neutrophils. In asbestosis, it is the asbestos fibre itself which when ingested by the macrophage activates it to release the neutrophil chemotactic factor.

The neutrophil has a short-life and must be continually attracted into the lung from the blood. It can produce and release many mediators of inflammation and fibrosis and one would expect patients with a high neutrophil count to do badly compared to those with a low count. This has recently been confirmed by a number of workers. Work over the last two years at the Brompton Hospital has shown:-

1. The volume of fluid recovered from patients with CFA is lower than controls.

2. There is a significant increase in neutrophils and eosinophils in CFA.

3. The cell counts in BAL fluid correlate with those extracted from open lung biopsy specimens.

4. Those with a predominant lymphocyte response respond to corticosteroids and tend to maintain this improvement over a six month period.

5. Those with an increase in neutrophils and eosinophils often fail to respond to steroids and may deteriorate over a six month period.

Alveolitis with increased lymphocytes

The cell consists of macrophages and lymphocytes without
neutrophils. The lymphocyte count is usually above 15% and the
increase has been shown to be mainly T lymphocytes.
Sarcoidosis is the commonest alveolitis in this group and a
typical BAL count in this condition would be: 50% macrophages,
40% lymphocytes with T lymphocytes being 35% of all cells
recovered (normally less than 7%). The T cells consist of more
helper cells and less suppressor cells than normal and are
activated releasing lymphokines such as monocyte chemotactic
factor (this attracts monocytes into the alveoli) and migration
inhibition factor (this immobilises the cells). Both these
factors are important in granuloma formation.

Gallium^{67} Scanning

This cyclotron-produced isotope has a half life of 72 hours
and when administered intravenously as Gallium^{67} citrate
localises in areas of inflammation. In active interstitial
lung disease, it is mainly taken up by activated macrophages.
The isotope is given intravenously and the patient scanned two
days later. To allow quantification of the uptake, the
"gallium^{67} index" has been developed (11).

1. Using a posterior scan the percentage of lung field for
    each area of uptake is measured.

2. The intensity of uptake is graded from 0 - 4.
    0 = background uptake
    4 = or is greater than the uptake of the liver.

3. The texture of uptake is weighted as either 0.5 for patchy
    uptake or 1 for diffuse uptake.

Thus an area of 15% of the lung field with an intensity
grade of 2 and diffuse (1) uptake contributes (15 x 2 x 1) = 30
units to the Gallium^{67} Index. This index correlates well with
the intensity of the alveolitis (as measured by lung biopsy and
BAL) in CFA and sarcoidosis. The radiation dose is
approximately 1 rad which is equivalent to a Barium enema
examination. It can therefore be repeated to assess the
intensity of alveolitis during therapy. In sarcoidosis, uptake
in hilar nodes occurs and must be differentiated from that in
the lung. Inflammatory airway disease does not give a positive
scan as the total mass of the airways is not great enough.

In CFA, there is a good correlation of the Gallium[67] index with the proportion of neutrophils in BAL. It has also been shown that activated T cells accumulate more Gallium[67] and normal lymphocytes. Gallium[67] scanning can therefore be used to stage CFA and allow an estimate of its prognosis.

BAL and Gallium[67] scanning are complementary. Some patients cannot be lavaged through either general contraindications or the presence of inflammatory airway disease. Gallium[67] scanning can be carried out in all patients and repeated. BAL is, however, more sensitive than Gallium[67] scanning and gives more information about the cell type.

Conclusion

The follow up of patients with CFA must be with techniques that allow an estimate not only of the overall state of the lung but of the intensity of the alveolitis.

Sympoms, physical signs, radiology and pulmonary function will be used to monitor the former and BAL and/or Gallium[67] scanning to monitor the latter. Hopefully this approach will allow more rational and effective control of drug therapy.

REFERENCES

1. Klassen, K. P., Anylan, A. J. and Curtis, G. M., (1949)
   Biopsy of diffuse pulmonary lesions.
   Archives of Surgery, 59, 694-704.

2. Knight, R. K. (1981).
   Bronchoscopy and other biopsy techniques.
   In: Thoracic Medicine (Butterworths), 189-219.

3. Steel, S. J. and Winstanley, D. P. (1967).
   Trephine biopsy for diffuse lung lesions.
   British Medical Journal, 3, 30-32.

4. Wright, P. H., Heard, B. E., Steel, S. J., Turner-Warwick, M. T. (1981).
   Cryptogenic fibrosing alveolitis: assessment by graded trephine lung biopsy histology compared with clinical, radiographic, and physiological features.
   British Journal of Diseases of the Chest, 75, 61-70.

5. Carrington, C. B., Gaensler, D. A., Coutu, R. E.,
     Fitzgerald, N.X., Gupta, R. G. (1978).
     Natural history and treated cause of usual and
     desquamative interstitial pneumonia.
     The New England Journal of Medicine, 298, 801-809.

6. Turner-Warwick, M., Burrows, B., Johnson, A. (1980).
     Cryptogenic fibrosing alveolitis: response to
     corticosteroid therapy and its effect on survival.
     Thorax, 35, 593-599.

7. Turner-Warwick, M., Burrows, B., Johnson, A. (1980).
     Cryptogenic fibrosing alveolitis: clinical features
     and their influence on survival.
     Thorax, 35, 171-180.

8. Shirai, F., Kudoh, S., Shibuya, A., Sada, K., Mikami, R.
     (1981).
     Crackles in asbestos workers: auscultation and lung
     sound analysis.
     British Journal of Diseases of the Chest, 75, 386-396.

9. UICC Committee. (1970).
     Cincinnati classification of the radiographic
     appearances of pneumoconiosis: a co-operative study by
     the UICC Committee.
     Chest, 58, 57-67.

10. Cole, P., Turton, C., Lanyon, H., Collins, J. (1980).
     Bronchoalveolar lavage for the preparation of free lung
     cells: techniques and complications.
     British Journal of Diseases of the Chest, 74, 273-278.

11. Line, B. R., Fulmer, J. D., Reynolds, H. Y., Roberts, W. C.,
     Jones, A. E., Harris, E. K., Crystal, R. G. (1978).
     Gallium-67 citrate scanning in the staging of
     idiopathic pulmonary fibrosis: correlation with
     physiologic and morphologic features and
     bronchoalveolar lavage.
     American Review of Respiratory Disease, 118, 355-365.

DISCUSSION

SPEAKER: KNIGHT   **CHAIRMAN: CUMMING**

CHAIRMAN: The paper is open for discussion

FABRI: First of all I would like to ask whether you don't
                 think it is advisable to perform an open pulmonary
                 biopsy in the area positive on a Gallium 67 scan
                 rather than selecting two areas at random.
                 Besides it seems to me to be a very good thing to
                 perform a BAL in patients with an oxygen partial
                 pressure less than 70 mm mercury; since we are
                 almost always dealing with very severe cases, we
                 perform these techniques even on patients with an
                 hypoxaemic level averaging 60 mm mercury,
                 obviously giving oxygen at the same time.

KNIGHT·   I didn't make it clear, the PO_2 that I was talking
                 about was the PO_2 that we could obtain with
                 supplement oxygen, not the PO_2 at rest breathing
                 air. I agree there are certain patients one can
                 safely lavage even if you have very low PO2, this
                 is meant to be an overall guide, and I think that
                 below that PO2 you have to assess individual
                 patients as to whether you are going to submit
                 them to Gallium 67 scanning on its own or lavage
                 as well. I think some of these techniques will
                 obviously depend on local expertise. As to
                 whether one should biopsy just areas that show up
                 on Gallium[67] scanning, hopefully this will give
                 you an index of the area of the activities of the
                 alveolitis macroscopically, I think it is probably
                 a good idea. Perhaps one should biopsy still,
                 though, an area that shows good uptake and an area
                 that shows poor uptake to give a better idea of
                 the overall disease.

ROSSI: At the beginning of your talk you said that one
                 can differentiate between a more cellular biopsy
                 and a more fibrotic biopsy, and that as in DIP and
                 UIP there is a correlation with prognosis you
                 assume that from the biopsy you can make a
                 statement about the prognosis. In IPF you can
                 state from a biopsy or from a lavage or from a

Gallium^{67} scan whether that patient has an active disease, but not predict the prognosis. You can assume that the patient with an active disease can be treated, but that's not correlated with the prognosis of the patient.

KNIGHT: I agree, hopefully in a long-term follow-up of these patients it will be, but I agree at the present it doesn't.

ROSSI: Second point. At a certain point, speaking about response to treatment in IPF you said that people with a higher level of lymphocytes in IPF or cryptogenic fibrosing alveolitis respond badly to steroids. My question is: why do people with cryptogenic fibrotic alveolitis have high lymphocyte levels in lavage fluid?

KNIGHT: I don't know, I don't think anyone knows. It would also be nice to know why they have high levels of eosinophils.

ROSSI: They do not have increased lymphocyte count in the lavage. I never saw a patient with IPF with high lymphocyte count.

KNIGHT: All the patients with cryptogenic fibrotic alveolitis have overall high neutrophil counts but in that population there are some who have higher lymphocyte counts than others and I was referring to those patients who have a higher lymphocyte count than others in the same population.

ROSSI: The problem is that in our experience the lymphocytes are in the interstitium, they are not in the alveoli. I mean, we have a higher percentage of cells producing immuno-globulins but the percentage of lymphocytes in the lavage is normal. Another thing, you said that gallium uptake in interstitial lung disease is done by macrophages. This is true but in IPF a lot of gallium is uptaken by neotrophils.

KNIGHT: I said it is taken up by active cells, I don't know anything about this, perhaps Rossi does.

ROSSI: Well I can answer indirectly, not directly, that question. What we know is that immune-complexes stimulate macrophages to pick up Gallium^{67} as a phagocytic response, but the exact mechanism I think is still unknown.

LEE: Have you correlated the lung washing cells and the Gallium scan? If you do your washing of the lung immediately after scan or at an appropriate moment, have you looked at the activity of the cells you get? That's the first thing.

KNIGHT: We haven't but they have, Dr. Rossi has.

LEE: The second thing is: in addition to studying particular neutrophils and lymphocytes, what about the immune-complexes that you may look at in the washings and other aspects, shall we say, transferring as an indication of porosity of the structure, as an indication of activity.

KNIGHT: Well, I don't know the answer, a tremendous amount of work that can be done has been done on these things, one of the problems is of quantifying it because of the dilution factor involved. What do you relate it to basically? What do you use as a marker? That hasn't been answered. There is a tremendous amount of work going on, looking at all sorts of things in alveolar fluid, but I can't give any positive answer I am afraid.

DENISON: Earlier on in your talk you said that one of the limitations of BAL lies in the difficulties we experience in understanding the dilution extent of cells and all the other alveolar molecular components. Don't you think this snag might be dodged by comparing the values of albumin in the patient's blood with the dilution one gets in the lavage? In fact the albumin molecule is small enough freely to penetrate the barrier between capillary and pulmonary alveolus. By doing so, in fact, you can correlate the level of the enzyme even in broncho-alveolar lavage, not only in the blood, and this appears to be more reliable.

KNIGHT: Some workers have tried it, I can see the theory behind it but I still don't know how you know that the albumin is present in the alveolar fluid in the same concentration that it is in the blood in disease states. That is what we were assuming, that may be true but I do not think it has been shown to be true.

PATHOLOGICAL EFFECTS OF DRUGS ON THE LUNG

B. Corrin

Cardiothoracic Institute

Brompton Hospital, London

It is estimated that 5% of all hospital admissions are due to drugs, that 10-18% of inpatients experience a drug reaction and that 3% of hospital deaths may be drug related (1,2). The lungs are often involved in these adverse reactions. Hutchinson et al (3) have described a potentially useful scheme for the operational assessment of whether or not a particular clinical manifestation represents an adverse drug reaction. This considers previous experience with the drug, alternative etiological candidates, the timing of events, drug levels, the effect of withdrawing the drug and rechallenge with the drug. The mechanism of a drug reaction may be based on overdosage, intolerence, a side effect, a secondary effect, hypersensitivity or idiosyncrasy (4). Drug reactions may be further classified according to the type of drug (analgesic, antibiotic, chemo-therapeutic agent, hormone, vasoactive agent, etc) or the pattern of disease. The last method is adopted here. Drugs may cause the following adverse pulmonary reactions:

(1)   Central depression of respiration

(2)   Broncho-constriction

(3)   Allergic bronchopulmonary effects
        (a) asthma
        (b) eosinophilic pneumonia
        (c) granulomatous alveolitis

(4)   Pleural disease
           (a) alone
           (b) together with pulmonary disease and possibly other
           manifestitations of systemic lupus erythematosus.

(5)   Alveolar injury
           (a) acute (diffuse alveolar damage)
           (b) chronic interstitial pneumonia and interstitial
           fibrosis

(6)   Alveolar histiocytosis/lipoproteinosis

(7)   Aspiration lesions

(8)   Pulmonary vascular disease
           (a) hypertension
           (b) haemorrhage
           (c) thromboembolism
           (d) drug embolism

(9)   Opportunistic infections

(10)   Metastatic calcification

(11)   Carcinoma

        Central depression of respiration occurs as a side
effect of barbiturates, morphine and its derivatives, and even
mild sedatives such as diazapam, and may be particularly
troublesome in patients suffering from chronic obstructive lung
disease. Ventilation in such patients may be largely dependent
on hypoxic respiratory drive and oxygen treatment may therefore
also have an adverse effect on respiration.

        Bronchoconstriction, as well as being mediated as an
allergic response, may occur as a pharmacological side effect
of several drugs given for non-respiratory diseases. These are
the cholinergic and β -adrenergic blocking drugs. Asthmatic
subjects are particularly susceptible. Whether aspirin induced
asthma is a pharmacological or allergic manifestation is
uncertain. Allergic bronchoconstriction forms part of
generalised anaphylactic reactions induced by vaccines and
antisera and as a local response to several drugs (Table 1).

        Eosinophilic pneumonia is the pathological basis of
pulmonary eosinophilia, which is defined as radiological chest
opacities associated with blood eosinophilia (5). Eosinophilic
pneumonia is characterised by eosinophils filling the alveoli.
There is generally also interstitial eosinophilic infiltrate

TABLE 1 – Drugs Causing Allergic Bronchopulmonary Disease

Asthma	Aspirin
	Antisera
	Penicillin
	Iodine (contrast media)
	Monoamine oxidase inhibitors
	Iron dextran
Eosinophilic pneumonia	Nitrofurantoin
	Para-amino salicylic acid
	Sulphasalazine
	Imipramine
	Phenylbutazone
	Gold
	Aspirin
	Penicillin
Allergic alveolitis	Pituitary snuff

but the hallmark of the condition is the intra-alveolar exudate
of eosinophils which is accompanied by varying numbers of
macrophages Poorly formed granulamata and a mild angiitis may
also be seen (6, 7). Causes of eosinophilic pneumonia other
than drugs include parasitic worms and fungi but many cases are
"cryptogenic".

Granulomatous alveolitis as an adverse drug reaction is
best exemplified by the extrinsic allergic alveolitis of
pituitary snuff takers. The granulomas are small, poorly
formed and non-necrotising. They are accompanied by a non-
specific interstitial chronic alveolitis and interstitial
fibrosis, most marked about the bronchioles.

Pleural disease manifest as effusion, chronic
inflammation or fibrosis may exist by itself or be associated
with chronic interstitial pneumonia, fibrosis and sometimes
serological evidence of systemic lupus erythematsus.
Methysergide is notable for the production of a pleural
effusion, pleural fibrosis and sometimes pulmonary fibrosis
whilst many drugs are associated with the development of a
syndrome resembling systemic lupus erythematosus (Table 2).
Whether these drugs are directly responsible for the syndrome
or merely promote the development of latent natural disease is
uncertain.

Diffuse alveolar damage represents an acute cytotoxic
injury produced by many noxious agents (8). Its clinical
manifestation is the adult respiratory distress syndrome.
Agents as diverse as viruses, X-rays, inhaled and ingested
chemicals (including certain drugs) and shock all strike at the
delicate cytoplasmic processes of epithelial and endothelial
cells which line the alveoli and the alveolar capillaries
respectively. The necrotic epithelium, together with
proteinaceous exudates from the blood, form hyaline membranes,
whilst damage to granular pneumocytes results in deficient
surfactant and hence collapse of the lungs. Radiologically
there is a "white-out" of the lung fields, often with a
"negative bronchogram". The pathological, clinical and
radiological features are very similar to those found in the
infantile respiratory distress syndrome. Survival is
associated with interstitial cellular proliferation, chiefly
lymphocytes and fibroblasts. With some drugs the onset of
disease is gradual and in these patients chronic interstitial
pneumonia and fibrosis develop without the severe acute
features of diffuse alveolar damage. Cytotoxic drugs such as
busulphan, methotrexate, cyclophosphamide and bleomycin are
particularly liable to produce acute or chronic forms of
alveolar damage The bizarre cells with large pyknotic nuclei

TABLE 2 — Drugs Causing Systemic Lupus Erythematosus

Hydralazine	Phenylbutazone
Procainamide	Thiouracil
Hydantoin	Reserpine
Sulphonamides	Thiazides
Penicillin	Methyldopa
Streptomycin	Digitalis
Isoniazid	Gold
Tetracycline	Aminosalicylic acid

TABLE 3 — Drugs Causing Acute Alveolar Injury and Chronic Interstitial Fibrosis

Busulphan	Azathioprine
Chlorambucil	Practolol
Methotrexate	Pindolol
Bleomycin	Gold
Cyclophosphamide	

described in busulphan lung (9) are seen with other cytotoxic
drugs and probably represent alveolar epithelial cells
attempting to regenerate in the presence of an antimitotic
drug. Drugs causing cytotoxic damage to the lungs are listed
in table 3. Hexamethonium, an obsolete ganglion blocking agent,
is not included as the fibrosis described may be the result of
prolonged cardiac failure and uraemia. Similarly
penicillamine, incriminated in the development of both diffuse
alveolitis and bronchiolitis obliterans, is also excluded
because these changes could well be due to the underlying
rheumatoid disease for which the penicillamine is administered.

    Histiocytosis was noted in rat lungs exposed to very
high levels of the antidepressant drug iprindole (10) and
similar experimental findings have been reported with the
anoretic drug chlorphentamine (11). The histiocytes (alveolar
macrophages) accumulate in response to lipidic bodies extruded
from the alveolar epithelium. The development of the lipid
represents a degenerative process (fatty change) falling short
of cell necrosis. The lipid represents the residual bodies of
focal cytoplasmic degradation i.e. that component of damaged
cell organelles which resists lysosomal digestion during
autophagocytosis. On being ingested by the alveolar
macrophages they prove just as indigestible and result in large
foam cells filling the alveoli (endogenous lipid pneumonia).
When the macrophages eventually disintegrate, compaction of the
lipid results in an appearance identical to that seen in human
alveolar lipoproteinosis, although here the mechanism is
probably based on hypersecretion of surfactant, as in the
silica-dusted rat (12). The drug levels used to produce this
interesting experimental finding are considerably in excess of
those used clinically. Such changes have not been reported in
patients receiving these drugs. In childhood however, alveolar
lipoproteinosis is often associated with lymphoproliferative
disorders and the relative roles of the primary disease,
cytotoxic drugs and opportunistic infections have not yet been
fully clarified.

    Aspiration lesions are generally those of exogenous
lipid pneumonia which results from the unintentional aspiration
of liquid paraffin or oily nose drops. Similar lesions are
sometimes seen with oil-based bronchography media. The
exogenous lipid forms globules which when dissolved out in
processing may be mistaken in paraffin sections for air-filled
alveoli. Foreign body giant cells lining such spaces give the
clue to the correct diagnosis. Often the pathologist is the
first to suspect the true diagnosis, this when a "paraffinoma"

is excised in the belief that the patient has lung cancer or at necropsy in a patient dying of "pneumonia" (13).

Pulmonary hypertension in many Swiss, Austrian and German patients in 1967 was probably due to the anorectic drug Aminorex (14). The pathology in these patients was identical to that of primary pulmonary hypertension and as it proved impossible to reproduce the condition in laboratory animals, proof that Aminorex was to blame is lacking. There is however very strong epidemiological evidence for this. The muscular pulmonary arteries showed medial hypertrophy and concentric cellular intimal thickening with the development of complex dilatation (plexiform) lesions. Also, very small arteries which are usually not muscularised, developed a muscular media. Experimentally such changes are produced by other ingested chemicals, notably the pyrollizidine alkaloids derived from plants of **crotolaria** and **senecio** species (15). Fenfluramine, an anorectic drug chemically related to Aminorex, may also cause pulmonary hypertension (16).

Pulmonary haemorrhage and thromboembolism may result from interference with the clotting mechanism by anti-coagulants on the one hand and contraceptive drugs on the other. Drug embolism, or more correctly "filler" embolism, is the result of illicit drug usage in which compounds designed for oral use are injected intravenously to heighten their effects. Oral preparations consist largely of fillers such as talc or starch and this insoluble particulate matter accumulates in the pulmonary capillaries. It provokes a foreign body giant cell reaction, thrombosis and fibrosis and may cause pulmonary hypertension (17).

Opportunistic infections are a common pulmonary hazard in any patient receiving steroids, cytotoxic drugs or any other immunosuppressant. Viral, fungal and protozoal infections may all develop in the lungs of such patients and tissue reactions may be atypical. Pneumocystis carinii infection for example may elicit a granulomatous reaction or cause diffuse alveolar damage instead of the usual foamy alveolar exudate (18, 19).

Metastatic calcification may result from high doses of vitamin D, calcium and inorganic phosphate or excessive alkali intake. Finally, carcinoma may rarely be promoted by drugs. Arsenicals cause squamous metaplasia of the bronchi and occasionally squamous carcinoma, whilst peripheral scar cancers, usually adenocarcinomas, have developed in lungs showing fibrosis due to drugs such as busulphan.

REFERENCES

1. Shapiro, S., Slone, D., Lewis, G.P. & Jick, H.
    Fatal drug reactions among medical inpatients.
    Journal American Medical Association, 1971, 216, 467-
    472.

2. Friedman, G.D., Collen, M. F., Harris, L. E., van Brunt, E.
    E. & Davis, L. S.
    Experience in monitoring drug reactions in outpatients.
    Journal American Medical Association, 1971, 217, 567-
    572.

3. Hutchinson, T. A., Leventhal, J. M., Kramer, M. S., Karch,
    F. E., Lipman, A. G. & Feinstein, A. R,
    An algorithm for the operational assessment of adverse
    drug reactions. II Demonstration of reproducibility and
    validity. Journal American Medical Association, 1979,
    242, 633-638.

4. Rosenheim, M. L., Editor. Sensitivity reactions to drugs.
    (Co-edited by Moulton, R.) Blackwell, Oxford, p.1.1958.

5. Crofton, J. W., Livingstone, J. L., Oswald, N. C. &
    Roberts, A. T. M.
    Pulmonary oesinophilia.
    Thorax, 1972, 7, 1-35.

6. Taskinen, E., Tukiainen, P. & Sovijarvi, A. R. A.
    Nitrofurantoin-induced alterations in pulmonary tissue.
    Acta path. microbiol. Scand. Sect. A. 1977, 85, 713-
    720.

7. Fiegenberg, D. S., Weiss, H. & Kirshman, H.
    Migratory pneumonia with eosinophilia.
    Arch. Intern. Med. 1967, 120, 85-88.

8. Katzenstein, A-L. A., Bloor, C. M. & Leibow, A. A.
    Diffuse alveolar damage - The role of oxygen, shock,
    and related factors.
    Amer. J. Path, 1976, 85, 210-224.

9. Heard, B. E. & Cooke, R. A.
    Busulphan lung.
    Thorax, 1968, 23, 187-193.

10. Vijeyaratnam, G. S. & Corrin, B.
     Fine structural alterations in the lungs of iprindole-
     treated rats.
     J. Path, 1974, 114, 233-239.

11. Heath, D., Smith, P. & Hasleton, P. S.
     Effects of chlorphentermine on the rat lung.
     Thorax, 1973, 28, 551-558.

12. Heppleston, A. G., Fletcher, K., Wyatt, I.
     Changes in the composition of the lung lipids and the
     "turnover" of dipalmitoyl lecithin in experimental
     alveolar lipoproteinosis induced by inhaled quartz.
     Brit. J. exp. Path., 1974, 384-395.

13. Salm, R. & Hughes, E. W.
     A case of chronic paraffin pneumonitis.
     Thorax, 1970, 25, 762-768.

14. Kay, J. M., Smith, P. & Heath, D.
     Aminorex and the pulmonary circulation.
     Thorax, 1971, 26, 262-270.

15. Kay, J. M. & Heath, D.
     Observations on the pulmonary arteries and heart weight of rats
     fed on **crotalaria spectabilis** seeds.
     J. Path. Bact. 1966, 92, 385-394.

16. Douglas, J. G., Munro, J. F., Kitchin, A. H., Muir, A. L. &
     Proudfoot, A. T.
     Pulmonary hypertension and Fenfluramine.
     Brit. Med. J., 1981, 283, 881-883.

17. Waller, B. F., Brownlee, W. J. & Roberts, W. C.
     Self-induced pulmonary granulomatosis. A consequence
     of intravenous injection of drugs intended for oral
     use.
     Chest, 1980, 78, 90-94.

18. Weber, W. R., Askin, F. B. & Dehner, L. P.
     Lung biopsy in pneumocystis carinii pneumonia.
     Am. J. Clin. Pathol. 1977, 67, 11-19.

19. Askin, F. B. & Katzenstein, A-L: A.
     Pneumocystis infection masquerading as diffuse alveolar
     damage.
     Chest, 1981, 79, 420-422.

DISCUSSION

SPEAKER: CORRIN   **CHAIRMAN: CUMMING**

CHAIRMAN: The paper is open for discussion.

PRICE: Mr. Chairman, I just want to make some very brief
                 points. First of all there is no question that,
                 as you mentioned yesterday, busulphan and
                 bleomycin specifically cause lung damage
                 irrespective of the way that you give them, but I
                 would like to point out that as far as the other
                 drugs are concerned, drugs like cyclophosphamide
                 and methotrexate, the toxicity to the lung is not
                 only a function of the drug itself but also of the
                 way in which it is given and if it is given
                 appropriately, these drugs very rarely cause lung
                 damage. Secondly, just for information and I
                 definitely don't want to be thought to be an
                 authority on this subject, but since it has been
                 raised, the Kaposi's sarcoma in male homosexuals
                 in New York, San Francisco and Atlanta is thought
                 now, as you indicated, not to be due to virus
                 infections, but due to what the Americans call
                 'recreational drugs'.   I am told that some of
                 these homosexuals have something like 60 partners
                 a week and among the drugs they take to keep them
                 going is amylnitrite and this is thought to
                 depress the immune system and so favour the
                 development of a sarcoma.

CORRIN: Thank you Len.

LEE: I suppose, like most of us, I have used aspirin
                 for most of my professional life but I have never
                 seen aspirin induced asthma. I presume this drug
                 effect is really very rare.

CORRIN: I guess it is.

DRUGS ACTING ON RESPIRATORY MUSCLES

Henry Gautier

Laboratoire de Physiologie Respiratoire

Faculte de Medecine Saint Antoine, Paris, France

Many drugs may act on the skeletal muscles including the respiratory muscles. Involvement of the respiratory muscles is of interest as it may change the pulmonary functions and ultimately the gas exchange. This review will not consider all the drugs which may affect the skeletal muscles but will focus on several particular points which may be of interest to the respiratory physiologist or the clinician.

Most of the drugs which may affect breathing will not act directly on the muscle fibre itself or on the muscle end-plate but rather on the control of the respiratory activity through the medullary respiratory centre and the spinal cord and ultimately, on the respiratory muscles. Schematically, three groups of drugs will be considered: (a) drugs which provoke an impairment of the respiratory function such as anaesthetic agents; (b) drugs which conversely provoke an increase in gas exchange such as respiratory analeptics and (c) drugs which may act at the periphery on the muscle end-plate like curare or on the muscle fibres like xanthines.

ANAESTHETIC AGENTS

Although it is generally agreed upon that anaesthesia may alter the respiratory function and gas exchange, the extent of the alterations are far to be fully understood. This can probably be explained by the various anaesthetic agents used, the common association with other drugs, the animal species under study and the methods used in the investigations. It is well known that anaesthesia may depress ventilation, even in humans, and modify the breathing pattern with a reduction in

263

tidal volume and an increase in breathing frequency (1). In addition, numerous studies have shown that anaesthesia provokes a slight but significant decrease of the functional residual capacity of the lungs in subjects lying supine in recumbent posture (2,3). It occurs on induction of anaesthesia and does not seem to be changed by the type, depth or duration of anaesthesia. This decrease in FRC has been observed using gas dilution methods and has also been confirmed using a body plethysmograph which includes measurement of gas trapped with airway closure (3). Several mechanisms have been proposed as potential cause for this FRC decrease; increase in central blood volume, gas trapping behind closed airways, atelectasis. More important are probably the findings which concern the changes in the mechanical properties of the chest wall and the lungs and the possibility that anaesthesia decreases the chest wall recoil was first raised in the study of Westbrook et al. (4). In this context, the findings concerning a cephalad shift of the diaphragm on induction of anaesthesia is of interest (5). The shift has been explained by a loss of tonic activity of the diaphragm which has been observed during halothane anaesthesia.

Another important finding which concerns the effect of anaesthesia on the respiratory muscles is the relative contribution of the rib cage and abdomen-diaphragm to tidal volume breathing. As early as 1858, John Snow wrote that the inhalation of chloroform leads to breathing sometimes only performed by the diaphragm while the intercostal muscles are paralysed (2). Observations of these alterations in the chest wall motion have been used to help gauge the depth of anaesthesia since Miller who described in 1925 the ascending respiratory paralysis under general anaesthesia (6). Recent studies have attempted to quantify the relative contribution of the rib cage and abdomen to breathing during general anaesthesia. During quiet breathing in the awake state, the mean rib cage contribution is less than abdomen-diaphragm contribution (15 to 43%). During anaesthesia with halothane, movements of the rib cage and abdomen are both depressed; this results in a much smaller tidal volume partly compensated by an increase in respiratory rate. In addition the contribution of the rib cage is much more depressed than that of the abdomen-diaphragm (7,8). The decrease of rib cage contribution is associated with a reduction of the EMG of inspiratory intercostal muscles (8). Evidence that alterations in chest wall function may lead to secondary changes in lung function has come from studies using chest wall restriction in conscious seated subjects. Changes in pulmonary functions like reduction in FRC, increased lung recoil pressure, decreased lung compliance which were observed during chest wall restriction

are similar to those seen with general anaesthesia and may
provoke alterations in gas distribution (9). At this point, it
must be emphasised that although the diaphragm is only a
respiratory muscle, intercostal muscles are both inspiratory
and postural muscles which stabilise the chest wall and prevent
inward movements during inspiration. Consequently, during
airways obstruction in non-intubated patients, a paradoxical
rib cage movement may be observed. This points out to the fact
that efficiency of the ventilatory function of the diaphragm
depends critically on the functions of the rib cage muscles
which must stabilise the rib cage against distortions induced
by diaphragmatic contractions. During anaesthesia, intercostal
mononeurones pool are more depressed than phrenic motoneurones
pool by an action of the anaesthesia at the brain stem and/or
at the spinal cord level. In a recent study (10), it was shown
that in subjects who were given intravenous morphine (0.15
mg/kg), a similar decrease in the contribution of the rib cage
was observed. The authors pointed out rightly that anaesthesia
or morphine given to patients with hyperinflations or increased
abdominal loads who rely heavily on intercostal muscles may
provoke a severe hypoventilation. It is worth mentioning that
during rapid eye movements sleep, the postural tonic activity
of all the intercostal muscles also disappear like in
anaesthesia.

RESPIRATORY ANALEPTICS

They are defined as drugs which are capable of restoring
depressed medullary functions. However, they are usually non
specific stimulants of the central nervous system capable of
causing convulsions at doses sometimes dangerously close to
those which cause respiratory stimulations.

They have been proposed for:

- post-anaesthetic respiratory depression.

- acute sedative drug overdosage although more conservative
measures are nowadays preferred.

- neonatal resuscitation especially when medication used during
delivery may have residual effects in the newborn.

-chronic respiratory insufficiency although use of respiratory
analeptics is much disputed except maybe in primary alveolar
hypoventilation.

-central respiratory regulation impairment such as Cheyne-
Stokes breathing, sleep apneas and Pickwickian syndrome.

-high altitude hypoventilation or chronic mountain sickness.

Although it has been considered in many studies, the efficiency of a drug as respiratory stimulant and its mechanism of action raise always many problems, not considering the untoward side effects: (a) experimental studies on ventilation are often performed on anaesthetised animals and obviously, the results cannot be simply extrapolated to patients; (b) animals are given air or oxygen to breathe although patients who would benefit from the drug generally show an hypoxia sometimes associated with hypercapnia; (c) even when drugs are evaluated in humans, this is done in subjects with normal respiratory pump (that of animals is supposed to be so) but the drugs will often be used in subjects with impaired respiratory mechanic with a low efficiency as far as gas exchanges are concerned; (d) there is also the choice of the parameters studied to account for the efficiency of the drug. Total ventilation is probably a good index during air or even oxygen breathing as patients are sometimes given oxygen to breathe. However, if a drug is likely to act through arterial chemoreceptors, its efficiency during oxygen breathing may be significantly reduced. Ventilatory response to hypercapnia or hypoxia (or both) is often studied but it would be hazardous to extrapolate the results to an hypoxic and eventually hypercapnic patient. It would probably seem better, although somewhat harder, to study the effects of the analeptic during mild exercise in patients with chronic respiratory insufficiency; (e) finally, the study of the ventilatory response to drugs in terms of tidal volume and breathing frequency should be more interesting than that of the overall ventilatory output in view of the fact that patients have often a respiratory pattern consisting of a low tidal volume and high breathing frequency. Ideally, a respiratory stimulant should be more efficient if it provokes an increase in tidal volume rather than in breathing frequency in order to improve gas exchange.

Among the many respiratory analeptics which have been proposed, several present some interesting characteristics (11). Xanthines, especially the caffeine, although known for more than a century may still present some interesting features; its effects have been attributed to a direct, though not clearly demonstrated action on the respiratory centres. Changes in ventilation are mostly caused by an increase in breathing frequency with no changes in tidal volume; this effect and the fact that they may cause a general excitation of the central nervous system explain that they are of little help in most patients. However, some xanthines, especially the theophylline is interesting as it can provoke a bronchodilation which may be useful in obstructive pulmonary disease. Also, as will be discussed below, xanthines may directly act on the muscle fibre and would be interesting in improving contractibility of the respiratory muscles and in overcoming the now popular respiratory muscle fatigue.

The mechanisms of action of doxapram have been relatively well worked out. Doxapram seems to act on arterial chemoreceptors and on brain stem respiratory centres. It provokes an increase in breathing frequency and sometimes in tidal volume but its overall value as a respiratory analeptic has often been questioned.

Finally, the most recent of respiratory analeptic, almitrine would mostly act on arterial chemoreceptors and as a consequence, its effects are usually significantly reduced by oxygen breathing. This drug may be of interest as it usually provokes an increase in tidal volume with little change in breathing frequency; on the other hand, it has been claimed that it can provoke an improvement in arterial blood gases without great changes in overall ventilation.

As far as analeptic properties are concerned, progesterone seems to be an interesting substance. Although it is known for a long time that a hyperventilation and lower PaCO2 is observed during pregnancy and during the luteal phase of the menstrual cycle, the ventilatory effects following the administration of the drug have been studied only recently. In normal subjects and in most patients with chronic obstructive pulmonary disease and CO2 retention, medroxyprogesterone provokes a hyperventilation due to a larger tidal volume and a significant decrease in PaCO2 and a blood and CSF alkalosis (12). What is intruiging about progesterone is that the mechanism of action is completely unknown and this is because, so far, no animals which could be used as an experimental model, seem to respond to progesterone as far as ventilation is concerned.

CURARE

When a drug which interferes with neuro-muscular transmission is used, muscles of limbs, neck and trunk are involved before intercostal muscles, and ultimately, the diaphragm is paralysed. Recovery of muscle activity usually occurs in the reverse order to that of their paralysis so that the diaphgragm is ordinarily the first to regain function. This observation is very important when one considers the use of relaxants during anaesthesia since the ideal blocking agent, when given in a convenient dosage, would act selectively on non-respiratory muscles and would relatively spare respiratory

muscles. This would allow muscle relaxation to be produced
without large impairment of spontaneous ventilation.

All the reports are in agreement and show that the hand-grip
strength is more depressed than the respiratory variables by
paralysing agents. Thus, when the hand-grip strength is
decreased by 80-90% the respiratory values are decreased by
only 20-50%. This is the case for the respiratory volumes
(except in some cases for the functional residual capacity),
maximal flows or pressures are also depressed. Also, weakness
seems to be greater for expiratory than for inspiratory muscles
(13). Another interesting finding concerns the effects of
curarisation which greatly prolongs the breath holding time.
The distressing sensation normally experienced during breath
holding is absent. This suggests that the distress of breath
holding arises in sensory structures stimulated by contraction
of the respiratory muscles and is not due to consciousness of
stimuli arising in the lungs (14). On the other hand, with a
partial curarisation in man with a resulting reduction of hand-
grip strength by 30-40%, the ventilatory response to exercise
is not surprisingly reduced, it is even increased by up to 50%.
Since the humoral factors were maintained constant in these
experiments, one has to postulate that the increase in
ventilation was caused by some nervous factors although the
origin of the factors, whether central or peripheral, remains
unknown (15).

The foregoing experiments were carried out in conscious
subjects and the differences observed between respiratory and
non-respiratory muscles may not be so apparent when the subject
is under anaesthesia during which the possible depressant
effect of many anaesthetic agents is added to the depressant
effect of the paralysing agent on the respiratory function.

The differences observed in response toward blocking agents
of respiratory and non-respiratory muscles remained to be
explained. Several hypotheses concerning differences in blood
flow, temperature and innervation have been put forward (16).
One has to consider all the chemical and electrical events
which take place in neuro-muscular transmission and which may
differ in these two groups of muscles.

Many drugs, especially the central nervous depressants, when
used in relatively high doses, may cause muscle relaxation with
possible respiratory impairments associated with sedative
effects by an action on muscle, spinal (namely reflexes) and
supra-spinal structures. This is especially the case with
drugs used to diminish muscle tone such as phenothiazine
derivatives, mephenisine, diazepam and meprobamate. When these

drugs are used in anaesthesiology in association with blocking agents, they may potentiate the effects of a peripheral muscle relaxant such as a paralysing agent (17).

AMINOPHYLLINE

As indicated above, some xanthines, especially the aminophylline seems to have a direct action on the muscle fibre. Thus, it has been shown in normal human subjects that the relationship between electrical activity and the pressure generated by the diaphragm could be changed with infusion of aminophylline: on average, at a given electrical activity, the pressure is increased by 15%. Furthermore, when diaphragmatic fatigue was produced by breathing against a load, the pressure generated by the diaphragm increased with aminophylline as compared with the pressure after identical fatigue runs without aminophylline (18). The precise mechanisms by which aminophylline enhances the contractibility of the diaphragm is not established, it may be by increasing cyclic-AMP, by acting on intracellular calcium or by blockade of adenosine receptors. These findings may have important patho-physiological and therapeutic implications, particularly in patients with chronic airflow obstruction or CO_2 retention, or both.

REFERENCES

1. Gautier, H. and Gaudy, J. H. (1978).
    Changes in ventilatory pattern induced by intravenous anaesthetic agents in human subjects.
    J. appl. Physiol., 45, 171-176.

2. Gelb, A. W., Southorn, P. and Rehder, K. (1981).
    Effect of general anaesthesia on respiratory function.
    Lung, 159, 187-198.

3. Hedenstierna, G., Lofstrom, B. and Lundh, R. (1981).
    Thoracic gas volume and chest-abdomen dimensions during anaesthesia and muscle paralysis.
    Anesthesiology, 55, 499-506.

4. Westbrook, P. R., Stubbs, S. E., Sessler, A. D., Rehder, K. and Hyatt, R. E. (1973).
    Effects of anaesthesia and muscle paralysis on respiratory mechanics in normal man.
    J. appl. Physiol., 34, 81-86.

5.   Froese, A. B. and Bryan, A. C. (1974).
     Effects of anaesthesia and paralysis on diaphragmatic
     mechanics in man.
     Anesthesiology, 41, 242–255.

6.   Miller, A. H. (1925).
     Ascending respiratory paralysis under general
     anaesthesia.
     J. Am. Med. Ass., 84, 201–202.

7.   Jones, J. G., Faithfull, D., Jordan, C. and Minty, B.
     (1979).
     Rib cage movement during halothane anaesthesia in man.
     Brit. J. Anaesth., 51, 399–407.

8.   Tusiewicz, K., Bryan, A. C. and Froese, A. B. (1977).
     Contributions of changing rib cage – diaphragm
     interactions to the ventilatory depression of
     halothane anaesthesia.
     Anesthesiology, 47, 327–337.

9.   Schmid, E. R. and Rehder, K. (1981).
     General anaesthesia and the chest wall.
     Anesthesiology, 55, 668–675.

10. Rigg, J. R. A. and Rondi, P. (1981).
     Changes in rib cage and diaphragm contribution to
     ventilation after morphine.
     Anesthesiology, 55, 507–514.

11. Gautier, H. and Bonora, M. (1982).
     Effects of hypoxia and respiratory stimulants in
     conscious intact and carotid-denervated cats.
     Bull. europ. Physiopath. Resp., 18, 565–582.

12. Skatrud, J. B., Dempsey, J. A., Bhansali, P. and Irvin, C.
     (1980).
     Determinants of chronic carbon dioxide retention and
     its correction in humans.
     J. Clin. Invest., 65, 813–821.

13.   Gal, T. J., and Arora, N. S. (1982).
       Respiratory mechanics in supine subjects during
       progressive partial curarization.
       J. appl. Physiol., 52, 57-63.

14.   Campbell, E. J. M., Freedman, S., Clark, T. J. H.,
       Robson, J. G. and Norman, J. (1967).
       The effect of muscular paralysis induced by
       tubocurarine on the duration and sensation of breath-
       holding.
       Clin. Sci., 32, 425-432.

15.   Asmussen, E., Johansen, S. H., Jorgensen, M. and
       Nielsen, M. (1965).
       On the nervous factors controlling respiration and
       circulation during exercise.
       Acta Physiol. Scand., 63, 343-350.

16.   Johansen, S. H., Jorgensen, M., and Molbech, S. (1964).
       Effect of tubocurarine on respiratory and non-
       respiratory muscle power in man.
       J. appl. Physiol., 19, 990-994.

17.   Gautier, H. and Vincent, J. (1981).
       Muscle relaxants and breathing.
       In: Respiratory Pharmacology, J. G. Widdicombe, Ed.,
       Pergamon Press, Chapter 16, pp. 335-342.

18.   Aubier, M., De Troyer, A., Sampson, M., Macklem, P. T.
       and Roussos, C. (1981).
       Aminophylline improves diaphragmatic contractibility.
       New Eng. J. Med., 305, 249-252.

DISCUSSION

SPEAKER: GAUTIER   **CHAIRMAN: CUMMING**

DENISON: I am not clear whether or not you explained the
               contradiction between the presumably radiographic
               findings on the position of the diaphragm in the
               anaesthetised patient that it appeared to be high,
               and the observations with the magnetometer which
               suggested that the great majority of the motion of
               breathing in the anaesthetised patient was
               abdominal. Can you explain the contradiction
               there? Why does the diaphragm rise in the chest
               when you anaesthetise and why does it make bigger
               excursions?

GAUTIER: Observations on the shift in the diaphragm
               position have been made using halothane
               anaesthesia and have been explained by a decrease
               in the tonic activity of the diaphragm (5). The
               increase in the contribution of diaphragm as
               compared to intercostal muscles has been seen with
               many anaesthetic agents including halothane. It
               may be that diaphragmatic fibres being more
               relaxed during anaesthesia will provoke greater
               volume displacement during inspiration for a given
               stimulation (Force/length relationship).

JENNETT: Concerning the increase in frequency under
               anaesthesia, you suggested some peripheral
               mechanical effects which might contribute to it
               but I was wondering about the possibility of a
               central action. Could it be that humans perhaps
               have some tonic inhibition of the frequency
               controller which is ruled out during anaesthesia?
               Could this possibility be related to the very
               common increase in respiratory frequency which
               occurs in comatose patients with brain damage? I
               wonder, I don't know, perhaps humans do know what
               frequency to breathe at when they are working on
               the brain only. Animals on the other hand perhaps
               have tonic drive to maintain high frequency in
               order to lose heat and they decrease the frequency
               when you anaesthetise them. I just raise these
               things as possibly relevant.

GAUTIER: Concerning the effects of anaesthesia at the
periphery, I like to remind you the work of Guz
(Clin. Sci. 1964, 27, 293) which showed that vagal
blockade in anaesthetised human subjects did not
modify the breathing pattern. This rules out any
important role for the anaesthesia at the
periphery in the genesis of the breathing pattern.
Now, concerning the central effects of
anaesthesia, we had recently the opportunity to
record breathing in patients recovering from
barbiturate overdose (Brit. J. Anaesth., 1982, 54,
1041). During the period of recovery, tidal
volume and minute ventilation increased whereas
frequency decreased. These modifications were the
reverse of those noted during the induction of
anaesthesia with barbiturates in man. In short, I
agree with you that anaesthetised human subjects
change their breathing pattern as comatose
patients with brain damage and I also think that
this is caused by direct alteration in the central
regulation of breathing pattern.